THE RHINOCEROS
AND THE MEGATHERIUM

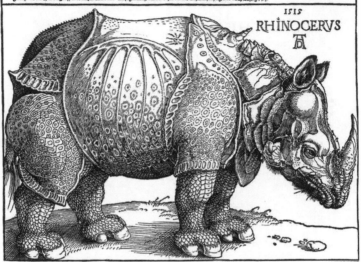

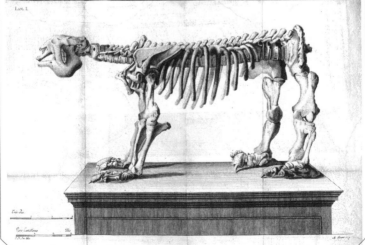

THE RHINOCEROS
AND THE MEGATHERIUM

AN ESSAY IN NATURAL HISTORY

Juan Pimentel

Translated by
PETER MASON

Harvard University Press

Cambridge, Massachusetts
London, England

2017

First published as *El Rinoceronte y el Megaterio: un ensayo de morfología histórica* by Abada Editores, S. L., copyright © 2010.

First printing

Library of Congress Cataloging-in-Publication Data

Names: Pimentel, Juan, author.
Title: The rhinoceros and the megatherium : an essay in natural history / Juan Pimentel ; translated by Peter Mason.
Other titles: Rinoceronte y el megaterio. English
Description: Cambridge, Massachusetts : Harvard University Press, 2017. | "First published as El Rinoceronte y el Megaterio: un ensayo de morfologia historica by Abada Editores, S. L., copyright (c) 2010."—Title page verso | Includes bibliographical references and index.
Identifiers: LCCN 2016019990 | ISBN 9780674737129 (alk. paper)
Subjects: LCSH: Rhinoceroses—Morphology—History. | Megatherium— Morphology—History. | Rhinoceroses in art. | Megatherium—In art. | Animals and history—Europe. | Zoology—Europe—History.
Classification: LCC QL737.U63 P5313 2017 | DDC 599.66/8—dc23 LC record available at https://lccn.loc.gov/2016019990

CONTENTS

THE RHINOCEROS
AND THE MEGATHERIUM

PROLOGUE

Analogy and Essay

In the "fantastic binomial," the words are not taken in their
daily meaning, but freed from the verbal chains that hold them
together on a daily basis. They are "estranged," "shifted," thrown
against one another in a constellation that has never been seen
before. Hence they are in the best possible conditions for gener-
ating a story.

Gianni Rodari, *The Grammar of Fantasy*

What follows is a historical essay with a tentative and
slightly provocative character. It proceeds rather like one
of those experiments of old in which assayers exposed materials
to strange conditions simply to see what happened. The project
is born of curiosity, although it must be confessed that it also
stems from a desire to play with ideas and forms. An essay is often
entertaining, a diversion, and I hope that this one entertains by
taking pleasure in discovering and telling stories, a task held in
scant respect by too many professional historians. Our task, they
say, is a serious one—as though pleasure were not serious too.

Also against the grain, in the face of a powerful emphasis on
the role of historical memory, we assert here the importance
of the imagination in the production of knowledge, including
historical knowledge.[1] I could deliberate at length on the status

and nature of the facts with which a historian works, as well as on the instruments and tactics that we use to create plausible accounts. But I shall try to be brief (after all, I have just promised the reader entertainment and, besides, I am not the right person to mount the defense that the historical imagination deserves).

What is certain is that even the most solid facts ("I was there," / "It happened to me," "I was present"—for of course credibility always starts with the first person) form part of a retrospective vision. They are described, worked into a narrative, and accompanied by argumentation—presented historically, in other words—with the aim of constructing a meaningful account that gives consistency to our identity. In short, they form a memory.

Does this mean that even our recollections are imagined? In a certain sense it does. Although the intellectual division of labor tends to draw lines of demarcation by throwing the imagination into the arms of fiction and the arts, there can be no doubt that imagination plays a part in all forms of representation and knowledge. One of the oldest of those forms, history, continually draws on the imagination. It does so when we read and try to visualize what we are being told. It does so when we compose accounts of places, events, individuals, and periods that we have never experienced in person and that we shall never experience (except perhaps with the help of science fiction).

No doubt the imagination helped Herodotus, the great emulator of Homer, when he narrated the Persian Wars and when he wrote about people with whom he had never been in contact. The same is true of Michelet when he put together that stylized vision of the Middle Ages that shaped our views for so long. Even

Karl Marx had to resort to the imagination to give shape to something that had never been conceived before: social classes. The pre-Socratic philosophers drew on it when they talked about atoms long before the electronic microscope, and the Cartesians used it in their descriptions of the whirling movements of parti-cles. The list could be extended to all those historians, scientists, and other producers of knowledge who imagine and represent the unprecedented, invisible, or hidden—things that they did not experience and could never have remembered, phenomena or entities that only emerge or assume form with the creation of a new instrument, question, or vantage point.

This essay focuses on two animals that were imagined without being seen, a pair of large quadrupeds transformed by the imagination into pictures that were reproduced and distrib-uted around the world. It is a history that engages more broadly with the relationship between image and imagination in the natural sciences of the modern era to explain how representa-tions of these two animals conferred a shape on the facts as they became benchmarks in the world of knowledge.

Images have a power to demonstrate that far exceeds that of theoretical pronouncements or chains of reasoning. It is no co-incidence that credibility, persuasion, the suspension of doubt, and access to truth are connected with the senses, whether we prefer touch (like Saint Thomas) or vision: seeing is believing. Imagination, the faculty with which we represent in images real or ideal things, has a crucial function in the making of evi-dence. Although it was not always so, visual rhetoric is more effective than the written or spoken word. The power of images in our world is incontestable.

One historian who has investigated the relations between science and art, Martin Kemp, noted the intersections that existed in Renaissance terminology for the imagination: *invention, fantasy, history, fable*.[2] The oldest and most complex of them, *fantasy*, derived from a Greek word equivalent to the Latin *visio*, alluded to the superior level of the imagination inasmuch as, according to Quintilian's *De Institutione Oratoria*, it is capable of inventing or producing things or images so vividly that they appear to be before our very eyes. The border between imagination and fantasy was always a delicate and problematic one. First, it seems clear that it is the dividing line that lies between production and reproduction, in other words, between the poetic and the mimetic, which is reflected in terms of genre with the distinction between history and fiction. However, dictionary definitions tell us that fantasy serves not only to idealize real things, but also to represent intangible things in a way that makes them accessible to the senses. Its aid is invaluable in helping us depict events distant in time or place. In that, it resembles the role of the imagination in the writing of history, and most certainly in recovering the two protagonists of this book, two extraordinary creatures from the remotest places: the other side of the world, and a past lost in the abyss of time.

For centuries there was no agreement on or regulated use of the term *imagination*. Of course, the term is found in the poetic and philosophical traditions of classical antiquity and the Middle Ages, where it was proverbially associated with the arts and with knowledge. The Scholastics thought that imagination and fantasy operated in one of the three cerebral ventricles, that of common sense (the others were reserved for the rational faculties of thought and for memory). That was where the impressions of the senses were gathered and processed by imagination and fantasy, the former to produce dreams and ingenious constructions, the latter to bring about free combinations. On the eve of the events narrated in this book, the architectural theoretician Filarete linked *fantasticare* with *pensare*, since "one of the aims of the artist is to investigate and *fantasticare* new things." This could be a definition of the man of science as we understand the term today. Francis Bacon consolidated the predictable division of labor: the imagination was assigned to poetry, reason to philosophy, and memory to history. John Locke did not attach too much credit to the imagination (he was aware that it could deceive in the quest for solid knowledge), and strongly linked it to the senses, especially sight and the world of images. Soon after, Joseph Addison similarly stressed the centrality of sight in his celebrated essay "The Pleasures of the Imagination." David Hume defined the imagination as a faculty for combining ideas in a natural way; if the combination was fanciful, it must have been the work of fantasy rather than imagination. Etienne Bonnot de Condillac placed imagination on the same footing as memory; they were both involved in the two basic psychological operations of reflection and analysis, in other words, the capacity to isolate, unite, abstract, compare, and relate ideas. Immanuel Kant, finally, also distinguished two kinds of imagination: a reproductive one, and a productive one that formed patterns and images.

Without too many pretentions, we confine ourselves here to stressing the importance of the imagination as an instrument of scientific and historical knowledge, one that is essential for the elaboration and shaping of facts, that is, to support, articulate, and convey reality. Paradoxically, perhaps it does so, as the Spanish philosopher María Zambrano suggested, "to remove ourselves from the overwhelming influence of facts, from the terrifying force of the immediate."[3]

However, it was Gianni Rodari, the Italian educator and writer who specialized in children's literature and whose work is regularly used in creative writing workshops, who provided me with the initial inspiration for this book. Taking up Novalis's complaint about the lack of a system of fantasy ("if we had a system of Fantasy as we have one of Logic, we would have discovered the art of invention"), Rodari wrote *La grammatica della fantasia* (1973).[4] It was there that I found the spark that gave rise to this book in the idea of the *fantastic binomial*, the combination and setting into motion of two objects or persons who are apparently unconnected. In creative writing, this exercise is used to stimulate the imagination and to generate a story. Exploring the hidden relations between two distant objects or persons, resting one's gaze on their profiles and characteristics, leads to unexpected and revealing stories that tell us a lot about the nature of each of those objects or persons. When viewed from an unexpected angle, a familiar landscape takes on a different aspect and meaning. Defamiliarization, making strange, recontextualization—many names can be used for it. Comparative or crossed history? I prefer the idea of the *fantastic binomial*.

In fact, our binomial is not actually so fantastic. It just looks a bit surprising at first sight. This essay sets out to assert the role of the imagination in the manufacture of scientific and historical facts. It does so by means of a classic mechanism, analogy, another Greek term that Aristotle, that great defender of the metaphor, used to refer to the correspondence between proportions. Analogy has been used throughout the history of science. We have Hippocratic medicine, with its theory of the four humors and their connections with the temperaments. We have the Neoplatonist tradition, which is entirely based on analogical thought: the correspondence between plants and stars, microcosm and macrocosm, the equivalence between geometry, music, and mathematics. But it would be wrong to suppose that, with the advent of modern science and the arrival of the literal (or algorithmic) reading of the great book of nature, analogy had had its day. Johannes Kepler was not the last of its subjects. Christiaan Huygens used a felicitous analogy with sound to formulate the undulatory theory of light. Isaac Newton deduced the shape of the earth by observing that of Jupiter and testing the hypothetical model of fluid spheroids. Charles Darwin managed to grasp the mechanism of natural selection by concentrating on how it was produced artificially among domesticated species. And James Clerk Maxwell's synthesis of physics arose from an ambitious analogy between optical, electrical, and magnetic phenomena.

There are many who have said, in one way or another, that science is the search for relations between apparently disparate phenomena or disciplines. Some spoke of hidden similarities in the style of Paracelsus, others of unity in diversity à la Alexander von Humboldt, the founder of biogeography. Even Peter Galison has offered us an Albert Einstein who knew how to extrapolate the problems of telegraph networks and the synchronization of railway clocks to conceive the theory of relativity.

So to use analogy to tackle a problem of knowledge is fairly unimaginative, totally lacking in fantasy, and really rather traditional. The double history has been a classic genre, from Plutarch's *Parallel Lives* right down to J. H. Elliott's *Richelieu and Olivares* (1987), to mention two illustrious examples. But here we are not proposing to compare a Greek and a Roman who faced similar moral dramas and problems, nor a cardinal and a favorite at court whose biographies were dominated by the administration of power in two absolute monarchies. Our fantastic binomial contrasts and links two quadrupeds almost three centuries apart and separated by different and yet comparable trajectories.

In a book that discusses questions like these and which has helped me to think about mine, Florike Egmond and Peter Mason pointed out that the weakness of this kind of research is that the phenomena compared may turn out to be disparate, heterogeneous, incommensurable.[5] Are the affinities and parallels between the rhinoceros and the Megatherium, the two characters of this book, the product of the imagination or of fantasy? Is the convergence or correspondence between the two histories reasonable or capricious? Is the analogy legitimate? The reader can judge.

As far as I am concerned, I can always claim that the two histories are mutually supporting, though this may not be fully evident to the reader until I become most explicit about the analogies late in the book. Each of the histories is a self-sufficient case, but when placed one after the other they offer an unusual opportunity for play, because in their diversity they nevertheless display points in common, and even similar problems resolved in different ways, which leads to comparison and contrast. The first history has even generated a novel, no less,[6] but as Rodari said, "It takes more than an electrical pole to produce a spark: it takes two of them." What the Italian writer said about words can be applied to our two protagonists: one in isolation only acts when it encounters the second, leaving well-trodden paths and discovering new significance. To cite Rodari again, "There is no life without struggle."

Although I do not want to reveal my argument prematurely, I will say that there is something symmetrical and mirrorlike about the double history of these big quadrupeds and their respective images. The first history is situated on the eve of the Scientific Revolution, the second at its close. My principal focus of attention has been how things from the past became known. I am a historian of science concerned with the modes of production of knowledge.

In this essay I have constructed an analogy related to the forms of two large vertebrates. It is an analogy situated in the dialectic between images and words, the fundamental rivalry in the history of culture, that "protracted struggle for dominance between pictorial and linguistic signs—according to W. J. T. Mitchell—each claiming for itself certain proprietary rights on a 'nature' to which only it has access."[7]

The search for isomorphisms between the two histories runs the risk of creating remote analogies. This is what they are called in synectics, a method of solving problems by using creative thinking and which assesses the correlation of the creative processes in art and science (by now some of my colleagues will think that I have joined Hare Krishna). In my defense, let me say that the play of analogies proposed here is not evident, but it is not absurd either. The method followed was to select and define what could be compared, to privilege certain patterns and configurations, and to leave aside the contexts[8]—in other words, to downplay whatever got in the way of perceiving the similarities in cases that are in so many other ways far apart.

The riskiest hypotheses are the most creative, but, as with scientific and artistic representations of nature, historians tend to look for a reasonable resemblance between what they narrate and the past. As an intellectual exercise, history does not copy reality; it broadens it. It does not transcribe; it recreates. But in the process its objects should not be allowed to become unrecognizable, absurd, or simply incredible. Any historian of science would jump up at this point to declare that the plausible, the logical, and the credible are historical categories that depend on the context and on the regime of beliefs and truths shared by the author and the readers of a historical or scientific text. Anyone else will say quite plainly that the pact of trust between the narrator of a history and the audience is connected with common sense and the feeling of reliability conveyed by a voice.

How did I choose this theme and how did it develop? There are many rational ways to answer this question, and we historians like to be witty or imaginative when talking about other trajectories, let alone our own. In this sense, I could say that I am a cultural historian of science who first studied the relation between science and politics, then between science and literature, and that I have ended up involved in a question at whose core lies the relation between science and art. I could add that I have shifted my focus from travelers and the movements of individuals, practices, and scientific ideas to the circulation of objects and nonhuman subjects.

The two cases place the Iberian world at the center of globalization and the processes of circulation described. I was not really aware of this when I was working on the book, but today it seems to me to be one of its most interesting assets for English-speaking readers. The role of the Iberian nations in the making of the modern world and their contribution to the first globalization are historical facts that ought to go beyond both the nationalist versions of the imperial past and interpretations connected with the Black Legend or the marginal position of those nations within the context of Western culture. In fact, the history of science is one of those disciplines that teaches us how narrow and limited the nation is as a historical subject. It is thus a splendid antidote to nationalism, racism, and other forms of collective narcissism. In a natural but perhaps not accidental way, the two protagonists of this book were both put into circulation in Iberian scenarios. The first was brought to Lisbon from the East Indies; the second arrived in Madrid from the Indies in the West. From then on, through the intervention of many other actors from central Europe, Italy, France, and elsewhere, they were transformed into scientific objects of global stature. Certainly the importance of Dürer and Cuvier in our double history seems to indicate the primacy of German humanism and French Enlightenment to natural history. But far from the competitive and rather childish character of such approaches, the cosmopolitan and transnational character of such enterprises as the print, the traffic in exotic species and fossil bones, comparative anatomy, and paleontology soon emerges.

The epicenter of this book on the global history of science is the Iberian world, a cultural space dominated by exchange, contact zones, and hybridization. No doubt it was like many others, even if the fantasy of purity still haunts certain cultural and national imaginations. There was a time when science too represented itself as a pure activity: disinterested, universal, rational, delocalized, incorporeal. Steven Shapin has deconstructed these arguments.[9] Certainly, a large part of the history of science of the last thirty years has set out to demonstrate that science was never pure but human—only too human—what the guardians of purity would call contaminated. Ever since the influential study of dirt by Mary Douglas, we know about the inverse relation that many cultures establish between purity and danger, but also, as the poet Hölderlin declared, that "where danger grows, salvation grows too."[10] This is a book of travels and knowledge, a book that talks about dangerous journeys, risky translations, and miraculous metamorphoses. Fortunately, the rescues outweigh the shipwrecks.

PART ONE

The Armored Pachyderm

CHAPTER ONE

Itinerary

On land too there are routes through which water runs, and
others through which the wind passes.

Leonardo da Vinci

He must have arrived exhausted after spending more than
four months caged in the hold of a ship. He had been captured on the other side of the world and was now—May 20,
1515—being led through the Tagus estuary. The vessel was
called *Nossa Senhora da Ajuda (Our Lady of Succor)*, the virgin
to whom mariners entrust their safety before going to sea.

According to the medieval bestiaries, the unicorn too could
be caught by a virgin, its sensuality overcome in the maiden's
presence. Captivated in her lap, it was easy prey for hunters.
Right from the start, the story of the rhinoceros evokes that of
the unicorn and other fantasies of the flesh: desires that need
to be assuaged. The members of the crew would have been
grateful to their *Lady of Succor* for bringing them home safely.
They had just completed the return journey of the Carreira da
Índia, the famous trade route from Portugal to India. But the
rhinoceros had been wrenched from his home.

A monastery dedicated to the same Lady of Succor rises on
one of the tall hills that dominate the river mouth. Below, in

the water, stands the Tower of Belém, under construction at the time to serve military and fiscal purposes and, according to some accounts or legends, to commemorate the voyage of Vasco da Gama. It is a fine example of why states and empires are built: to remember or to imagine traditions, to make war, and to collect taxes.

He would have disembarked there at the fortress, which doubled as a customs office, controlling traffic and merchandise. His effigy can still be seen today in one of the limestone arabesques that grace the base of the Manueline turret. Beside the armillary spheres, knots of the House of Avis, crosses of the Order of Christ, and other ornamental decorations of this singular Mozarabic structure that seems to issue forth from the very waters of the Tagus, a gargoyle recalls his neck and round head surmounted by his legendary appendix, a prodigious natural feature, shaped like a figurehead at the prow of a ship but made to charge at something more substantial than the wind—the flesh of an enemy, a powerful bulwark like himself, a rival of his own size (Figure 1).

He had come a long way. In his own country they called him Ganda, the name for rhinoceros in most North Indian languages. Weight: around two tons. Size: almost two meters high and more than three long. He must have felt dizzy and confused. His eyesight was bad; it had never been good, and was certainly not now. But his sense of hearing was acute—so much the worse, as there is nothing more disturbing than to hear strange noises without being able to spot the danger, without even being able to make out what they mean. During those months aboard, Ganda must have heard many unfamiliar sounds. Now everything, human and nonhuman, seemed to belong to an alien world of strangeness, novelty, curiosity. For

Figure 1. Belém Tower and its gargoyle. Pen and ink drawings by Jorrín Montañés.

his captors it must have been a game, the game of the hunt and the gift, the interminable game of court society (which is nothing but the society of war, symbol, and art). For him the uprooting must have been a calvary. Our story has to start with this zoological Via Crucis, this memorable day in the life of the hoofed pachyderm.[1]

HE WAS USED to the element of water—the river—but Ganda had been obliged to cross oceans. He was entrusted to the Portuguese near the riverine port of Surat, in the Gulf of Cambay, in the kingdom of Gujarat (or Cambay, as it was called by the Portuguese), Northwest India. From there he was taken to Goa, the center of Portuguese operations, over a distance of more than five hundred kilometers. Soon afterwards he was conducted a similar distance farther south to the port of Cochin on the Malabar Coast (where Vasco da Gama was to die in December 1524). From Cochin, Ganda embarked with the Portugal-bound flotilla of three vessels aboard *Nossa Senhora da Ajuda*, captained by Francisco Pereira Countinho, at the beginning of January 1515. The rhinoceros was accompanied by Jaime Teixeira, one of the ambassadors of Afonso de Albuquerque, the governor responsible for his transport to Lisbon and who was to die at the end of the year. Also on board was an Indian called Oçem, his handler and guide, though the fame of the pachyderm has overshadowed any further details about this man's life.

They crossed the Indian Ocean and probably took the "inside" route between Madagascar and Africa, what would later be known as the "old line." They doubled around the Cape of Good Hope to make a landing at St. Helena, a further one on Terceira in the Azores, and finally reached Lisbon in May. Three ports of call on a 120-day voyage—just enough to maintain enough fresh water for drinking and to replenish Ganda's supply of grass.

He came from remote Goa in confinement, wearing chains and adorned with a costume. He was now a precious object. The cargo included other precious commodities: cinnamon, pepper, myrrh, sandalwood, aniseed, cloves, and aloe. The spices were to conserve foodstuffs and to improve their flavor; the others—perfumes and colorants to produce scents and colors, knowledge—affected other senses. Like them, the rhinoceros was a natural product, an object of exchange, and a highly valued item because of his connection with the senses, especially those of sight and touch. As an exotic fetish, Ganda represented the sensuality of the East in opposition to the rationality of the West.

We might also consider him a prisoner, or rather an exile, a *degredado,* one of those deported criminals whose death penalty had been commuted in exchange for the performance of certain functions in the colonies.[2] Ganda was a wild animal who had been domesticated and degraded to a status somewhere between human and nonhuman—a sentient being and an object at the same time, one more victim of war and sport. And like all the products of the acquisitive mentality of the West, he was experiencing what traffic means, because all travel, and especially a journey like this one, entails hardship; it is not a coincidence that the etymology of the word "travel" links it with travail, torture, and labor.[3]

Ganda had learned this lesson the hard way. There is no getting from A to B without effort and distress. His body really was made to resist, to resist and invest, to give and receive. In this

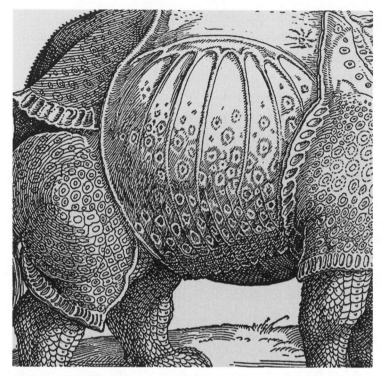

Figure 2. Detail of *Rhinocerus 1515*. Woodcut by Albrecht Dürer.

respect he resembles his captors, fascinated by the possibilities and jingling coins of commerce. Ganda, however, did not know the tricks of the symbolic trade. He was alien to the idols of the market and of sociability, those gods in the ascendancy after the Europeans had brushed with the New World and doubled the Cape of Good Hope. Who or what was Ganda? A suit of armor and a horn, made to resist and to attack. Only a body as capable of resistance as his could withstand such a brutal estrangement (Figure 2).

His arrival in Europe forms part of the gift economy. Ganda was a gift too. Uprooted from the jungle, he had just entered the circuit of symbolic exchange: give, receive, give in return.[4] In the year before he disembarked in Lisbon, he had been given by the Sultan of Cambay to Diogo Fernandes de Beja and Jaime Teixeira, the emissaries of the all-powerful governor of the Portuguese Indies, Afonso de Albuquerque.[5]

Was the Sultan of Cambay his first owner? As far as we know he was, but it would not be surprising if one of his vassals had previously presented it to him. The gift economy was globalized before the spice trade. The sultan is variously known as Modofar or Muzafar II or V, but all the Portuguese chroniclers (João de Barros, Diogo do Couto) agree in describing him as a slave to base passions at table and in bed, always surrounded by women (he had more than five hundred) and drugged with opium. This luxury and excess, however, are part of the stereotypical Western image of the Orient. So we can establish neither whether Muzafar/Modofar was the first captor of Ganda nor whether he was the second or the fifth of his dynasty nor whether he was licentious. But he needed to be so for the story, in which desire, prefiguration, and exoticism are recurring elements.

What is it exactly that is desired? We covet what is at hand, within view, but always desire what is remote. *Omnia nova placet*: everything new gives pleasure. Isabella d'Este, Marquise of Mantua, a serious collector and patron of the arts at the time, explained this better than anyone else when she defined her passion as "an unquenchable thirst for ancient things." She desired antiquity because it was newly discovered and remote in time. We desire what is new in time and remote in space in the same way. A Roman cameo, an Etruscan inscription, a rhinoceros—all that matters is that it is unfamiliar; opposites are attracted to one another, we are attracted to what is different. The West needs the Orient just as dying presupposes being born. Nothing could be more natural: the Orient plays a geographical role for humanism analogous to that played by Antiquity—a remote fascination finally recovered from the past.

The Renaissance searches for the Orient and even creates it. The East continues to be created and recreated through the whole early modern era down to the Enlightenment and the age of Romanticism, by which time it is consecrated as one of the great cultural artifacts generated by the West in its entire history.

The Orient does not exist as such. Decades ago Edward Said studied the mechanisms and interests of that motley disciplinary mosaic called orientalism.[6] How could Europeans be so unsophisticated as to include Pharaonic Egypt, Meiji Japan, and the Sassanid Empire in the same category? Only an obstinate and determined desire to find a cultural adversary, a reverse image, can have generated such a pronounced and violently condensed alterity.

The Orient does not exist, but we need it to exist. Europe needs it to such an extent that it even provokes the emergence of the other, fourth, unknown, and unexpected continent, America. But as soon as Europe has ensnared the Orient by the African route, it captures it, transports it, and puts it on display. Leaving metaphors aside, this is exactly what it did with Ganda.

However, what used to arrive by land, at least partly, now began to arrive by sea. Lisbon was supplanting Venice. The connection between the Mediterranean and the Indian Ocean is one of the landmarks of any history of the West, whether the old one of the age of discovery or the more modern one of globalization. Adam Smith stated that, together with the discovery of America, the doubling of the Cape of Good Hope had been the most important event in the history of humankind, and it was even more important for Voltaire. At the beginning of the sixteenth century, Venice was toying with the idea of opening up a maritime passage near the Suez. It was too early—or too late: Bartolomeu Dias and Vasco da Gama had already discovered a route around Africa.[7] The Portuguese had been marauding the Atlantic coast for around a century as they trafficked in gold and slaves in Guinea, making do with malagueta pepper until they finally discovered the route leading to genuine black pepper from Asia.

In a little less than fifteen years they pounced on the principal ports of the Sea of Arabia and the Gulf of Bengal. Afonso de Albuquerque (1453–1515), the official responsible for the deportation of our pachyderm and the intermediary between the Gujarat sultan and the Portuguese king, was one of the builders of the Portuguese colonial empire. He also directed the occupation of the island of Ormuz, turned Mozambique into the main landing stage on the *Carreira da Índia,* and even seized Malacca, thereby opening up the route to China.[8]

The Gulf of Cambay, Ganda's natural habitat, was impregnable. The Portuguese were hoping to establish two *feitorias* (trading posts) on the diminutive islands of Daman and Diu. In fact, the gift of the rhinoceros was connected with the operations on Diu. Albuquerque wanted to secure permission to erect a fortress on the island and sent his two emissaries, Beja and Teixeira, to Muzafar for that purpose. The gifts they were to present to the sultan included a gold dagger set with rubies, silver cups, and Persian brocades and collars. The sultan, however, refused to accede to Albuquerque's demands; in fact, when the emissaries returned to Goa with the negative reply, Albuquerque considered making war on him in retaliation. Fortunately, Muzafar knew that meekness was not among the qualities of the Portuguese governor. To palliate the wrath that he foresaw, he hit upon the happy idea of sending him a chest inlaid with mother-of-pearl with balustrades and ornaments for the king and a monstrous animal for Albuquerque himself. In the words of Gaspar Correia, whose *Legends of India* written some fifty years after the event is one of the earliest accounts of Portuguese rule in Asia, the creature in question was low in stature and had the hide and feet of an elephant, a long head like that of a pig, eyes near the muzzle, and a thick, stumpy, pointed horn on its nose. It lived on grasses, straw, and boiled rice. He also describes it as having a docile character.[9] Ganda was indeed tame, a suitable present to pacify a governor.

South of Cambay, on the Malabar coast, Albuquerque had directed the seizure of Goa, from then on the principal platform in India. Cochin, from where Ganda embarked, was much further south, near hostile Calicut. Without actually occupying India (something that only the British managed to achieve to a certain extent in the second half of the nineteenth century), the Portuguese monarchs considered themselves lords of the sea. Their thalassocracy had been established.

Statistics, however, deflate the imperial rhetoric. Almost three hundred vessels sailed the route to the Indies between 1500 and 1528 at the rate of around ten a year. The average dropped to half this in the following century. From a global perspective, the Portuguese presence in the East should be regarded as a minor phenomenon. Ming China, Safavid Persia, and the India of the Great Moghuls did not need to worry too much about the existence of a few commercial enclaves of these barbarians from the West, the *francos* or *rumies* (the Hindus regarded the Portuguese as members of the Frankish Empire or as Romans) whose only appreciable technological advantages were firearms and caravels.[10]

Ganda did not need to pay them much heed either. Like the Orientals for the Europeans and vice versa, all humans must have seemed pretty much alike in the eyes of a rhinoceros: Muzafar, Afonso de Albuquerque, Manuel I, pawns in a game of reciprocal gifts. Without knowing it, Ganda bore the whole weight and fantastic imagery of the Orient on his robust frame. When he arrived in Lisbon, the Orient—unfamiliar, dangerous, subjugated—had come to court.

THE HISTORY OF the capture and keeping in captivity of animals is a long one.[11] Virtually every culture practiced it. It is known that as early as the thirteenth dynasty Queen Hatasou had a zoological garden in Thebes. The Chinese sovereigns of the ninth century BC had another called the Garden of Intelligence because of its alleged divine creation. Marco Polo confirmed that the Great Khan also kept big cats in his palace and herbivores on his grounds. The Aztecs, and long before them the Assyrians, captured and kept wild animals for a variety of purposes: sacrifice, ritual, hunting, display. In India kings and dignitaries domesticated elephants, tigers, and lions.

The Greeks kept birds and monkeys in their residences, but it was Alexander the Great, influenced by the Persians, who incorporated the Oriental practice of domesticating wild animals and pachyderms, using them in processions, and displaying them as an expression of power. In Rome all these forms of captivity and sociability with other living beings were combined. Keeping exotic birds at home was a symbol of luxury. In the public arena it was not long before certain animals were let loose on the gladiators to gauge their ferocity. From the third century BC on, the Romans carried out a strange military ritual that consisted of sacrificing elephants and big cats captured from the enemy to avenge the losses suffered in battle. In the imperial era the presence of wild animals in circuses and military parades became frequent and contributed to making these kinds of events popular. As a consequence, the value of the animals increased and they became the object of traffic and diplomatic exchange.

Although these practices declined in the Middle Ages, they never disappeared completely, especially in Byzantium. The crusaders returned to Europe with certain customs that they had picked up from Islam. One of these was to hunt in the company of leopards, even going so far as to house them in their own residences. Charlemagne also received elephants and exotic birds from the princes of the East. One of his successors, the Holy Roman Emperor Frederick II, who was also king of Sicily, Germany, Cyprus, and even Jerusalem, was an outstanding figure in this field. Considered to be the Antichrist by the pope but held in high esteem by the Saracens (even though he made war on both of them), Frederick was one of those violent and cultivated persons with an interest in everything oriental. Apparently he spoke nine languages, at a time when it was not uncommon for monarchs to be illiterate. He founded the University of Naples in 1224 and became known as a living marvel, a *stupor mundi*. Like Sultan Muzafar, he had the reputation of being both lecherous and refined. His courtly virtues have been described as a sophisticated blend of intelligence, a lust for power, and an interest in everything new. He lacked neither pleasures of the flesh nor the desire to surround himself with exotic animals: camels, dromedaries, elephants, big cats, bears, gazelles, and even a giraffe. Guests at his wedding with the sister of Henry III of England could admire the imperial menagerie opposite the Cathedral of Worms. On that day he presented the English king, his new brother-in-law, with three leopards, or more probably three lions—a highly symbolic detail, given that the coat of arms of the Plantagenets featured not only a unicorn but also three lions. The gift of a sister and the gift of three lions in return sealed an alliance based on blood and violence. France was warned.

Lions or leopards, those three felines traveled to the Tower of London, where the English king kept several wild animals, a practice begun by his grandfather Henry I, the son of William the Conqueror. The first royal menagerie had been in the open countryside in Woodstock, near the city of Oxford, consisting of several exotic species: lions, leopards, lynxes, camels, and a strange creature that threw darts at its attackers, a porcupine. Soon afterwards, however, a menagerie was established in the Tower of London—the fortress that William the Conqueror had built and whose turret guarded the only bridge over the River Thames—to accommodate those extraordinary royal gifts, live animals in captivity, who were appreciated for their singularity and rarity.[12]

A procession of Frederick II's collection of animals passed through the streets of Ravenna, and on his journey to Verona he displayed an elephant, five leopards, and two dozen camels. The emperor knew how to impress his allies (or potential rivals) with these processions, ceremonial entries, and ostentatious visits in which the scenography and propaganda were so important.

He received a giraffe from an Egyptian sultan in exchange for a polar bear. The bear must have had a worse time of it in northern Africa than a rhinoceros in Europe.

It was not long before neither the Tower of London nor the behavior of the emperor was regarded as exceptional any more. Collecting live exotic animals was becoming a sign of distinction, the expression of symbolic power over distant territories and over nature itself.[13] Like other practices associated with natural history, such as the phenomenon of collections in cabinets of curiosities and the cultivation of species in botanical gardens, the menageries and collections of living animals began to proliferate in the princely courts of Italy from the end of the fourteenth century. These spaces continued to grow in size and diversity: open or closed, in interior annexes of the palaces themselves. They could be found from Naples in the south to Parma in the north. The Duke of Ferrara and the Duke of Calabria collected wild animals in their villas, and so did the Medici in Florence.

Lorenzo the Magnificent (1444–1492) is perhaps the most dazzling and best known of the latter. The story of how this grandson of Cosimo, a prosperous merchant, managed to join the ranks of the nobility and become the greatest prince of the Renaissance thanks, among other things, to the possession of a giraffe, has been delightfully told by Marina Belozerskaya.[14] The Florentine Republic kept some wild animals in cages in the mews adjacent to the Piazza della Signoria, particularly lions, the emblematic animal of the city and the creature most sought after as a diplomatic gift. As a boy, Lorenzo attended the public ceremonies held in honor of Pope Pius II and the Duke of Milan in Florence in 1459. The visit concluded with an animal combat in the Piazza della Signoria, with which Cosimo hoped to emulate the spectacles of the Roman circuses. In the event, however, the lions limited themselves to jumping onto a mechanically driven wooden giraffe and onto the backs of the horses, but without biting them, much less devouring them. They were too pampered for that. Their lack of ferocity was interpreted as a bad omen.

With time, Lorenzo grew to become a grand statesman and without a doubt the greatest patron of his day. He was a humanist, a friend of Poliziano, Ficino, and Pico, and a patron of Botticelli and Michelangelo. Many things can be said about this fascinating person, but never that he behaved like a tame lion. The man who united two souls in one body, one delicate and affable, the other violent and cruel, never lost sight of two principles: never give your enemies too many options; and know how to indulge your friends. In fact, they were the two sides of the same coin, a policy that consisted of impressing others and leaving them with an image or a mark (a trace, a scar) so they would remember who he was. In courtly society and public life, dominated as they were by gestures and propaganda, a prince— as Machiavelli reminded his readers—must not only possess certain qualities, but above all represent or feign his possession of them. Most people see what you present yourself to be; only a few really know you, but these do not dare to contradict the opinion of the majority (and if they do they will not often manage to change it, we might add). It is a devastating line of reasoning, as Machiavelli knew, an acute diagnosis of the dynamics of social life and perhaps the human soul.

After the conspiracy of the Pazzi and the invasion of Florence, Lorenzo deployed his best diplomatic skills and his great political talent to restore the prestige and break the isolation of the Florentine Republic. He traveled to Naples, where he headed a protracted diplomatic mission to convince the King of Naples, his fierce rival, to become his ally and thereby isolate the pope. This was soon followed by agreements with the French sovereign, a policy of marriage alliances, and even the election of one of his sons as pope.

Within that delicate marquetry of alliance, diplomatic missions, bank loans, and poisoned presents, that interminable repertoire of operations, deeds, and practices that include finance and war, the whole of politics and culture, that dense network of social relations whose leitmotiv is power, the thirst and the desire for power (in all of its meanings, in whatever variant), Lorenzo the Magnificent decided to obtain an animal that would elevate his social standing and distinguish him from the other princes of his day. This animal was a giraffe, the one that dominates the fresco crowning his room in the Palazzo Vecchio painted by Giorgio Vasari, in which Lorenzo is shown receiving the attentions, deference, and gifts of ambassadors from all over the world.

It was both a bold and a classic decision. Well aware of the tradition in which he wanted to ensconce himself, Lorenzo the Magnificent knew that other illustrious men had obtained one of these strange animals. They were so peculiar that they seemed to be a hybrid of a camel and a leopard, as reflected in the Latin *cameleopardus* and in the binomial nomenclature *giraffa camelopardalis*. And although the precedent of Frederick II was not a bad one—Frederick was king of five or six places, as hated by the pope as Lorenzo was, and when all was said and done a Holy Roman Emperor—it was Julius Caesar whom Lorenzo had in mind. After completing his campaign in Asia Minor and Egypt, Caesar had entered Rome in triumph accompanied by a procession of hundreds of lions, black panthers, elephants, monkeys, and a giraffe, the first to be seen in Rome, according to Pliny.[15] To the delight of the public, that giraffe was devoured by the lions in the middle of the spectacle. This left such a lasting impression that fifteen hundred years later a Florentine merchant and moneylender, Cosimo de Medici, the grandfather of Lorenzo, had a wooden giraffe on wheels constructed for his lions in order to represent the same scene, though the results, as we have seen, failed to impress. So from a very early age Lorenzo would have understood two things; that lions should not be overfed, and that the mirror in which he should reflect himself was the world of ancient Rome. Nothing could satisfy him more than to emulate or to be seen as a *Caesar redivivus*. His unstoppable rise from his patrician origins to a status in between human and divine inspired him. Lorenzo recognized himself in Julius Caesar, just as Caesar had wanted to be seen as a second Aeneas, from whom he even claimed descent. Heroes, ancestors, references: no man, however great, has failed to search for a figure from the past to identify with. In fact, it is those in quest of fame who tend to take a lively interest in how others obtained it before them.

Lorenzo the Magnificent managed to get his hands on a giraffe during a series of diplomatic operations and pacts with the Mamluk Sultan of Egypt, Qaitbay, aimed at strengthening their reciprocal trading relations and waging war on their enemies. Like Portugal, Florence was trying to take the place of Venice in commerce with the Orient. As for the Sultan of Egypt, he wanted support, arms, and the papal blessing to fight the Ottoman Empire. He obtained these thanks to the emissaries he dispatched to Italy in 1487. It was this mission that left the giraffe in the Piazza della Signoria after a grand entry that also left its mark on the city. Those Mamluks paraded with their animals and tents in an ostentatious display of brocades and muslins. They placed at Lorenzo's feet some Chinese ceramics and porcelain that had never been seen before and presented him with the large giraffe, a tribute fit for a Caesar. The procession resembled the coming of the Magi, an epiphany from the East.

In exchange for the giraffe, Qaitbar wanted Lorenzo to intercede to introduce him to Prince Cem, the unfortunate younger brother and rival of Sultan Bayezid II to the throne of Turkey. Exiled in Europe, Cem was held prisoner in a castle in France. He was thus a key figure in any attempt to overthrow the Ottomans and shift the balance of power in the Levant. Lorenzo wanted Queen Anne of France to accept the giraffe in exchange for the release of the coveted prisoner, but everything went wrong through a series of circumstances that can be passed over here but included the premature death of the giraffe. Apparently while eating in a wood in Florence one day in January, its head became entangled in some branches, and when the animal made an abrupt movement its neck was broken. The next year, however, Lorenzo achieved a different triumph when his son Giovanni was elected cardinal at the age of thirteen. It was 1489. Bartolomeu Dias had just returned to Lisbon after doubling the Cape of Good Hope.

To return to Ganda, one might say that Lorenzo the Magnificent interpreted correctly what antiquity dictated but only the Orient could provide, in the same way that what archaeological objects, canvases, or stones legitimized could perhaps only be fully achieved by a living being, a natural prodigy to complete the marvels of art. The possession and exhibition of exotic animals from the Orient affected such important courtly practices and actions as the need for distinction, the desire for nobility, the securing of a certain *sprezzatura*, and the wish to be surrounded by precious objects and creatures from afar. They were a highly appreciated diplomatic gift and could be used to seal agreements or even to unleash wars. Their death was regretted by those in close contact with them and was regarded as a bad omen.

A giraffe, an elephant, or a rhinoceros was not just an exotic item that would add an orientalizing touch to the decorative and fine arts. It also offered a tangible and organic counterpart to material wealth, a living and therefore more fragile display, more valuable than the most valuable work of art because of its ephemeral nature. To be able to keep an extraordinary living animal from a remote region was the supreme expression of power—power to encircle and encompass the world, to bring together what was different and unique in a collection, to lay up not only manmade but also natural treasures. This was beyond the ambitions of all but a very few.

WHEN GANDA ARRIVED in Lisbon, his new owner, King Manuel I, made two important decisions. The first was to organize a duel between the rhinoceros and an elephant in Lisbon, an episode to which we will return. After the summer, however, he decided to present the rhinoceros to the son of Lorenzo de Medici, who had become Pope Leo X. Ganda thus set out once more on a Mediterranean voyage. He had spent barely seven months in Lisbon.

The latter decision confirms the irresistible ascent of our pachyderm, who was gaining not only seafaring experience but also other human attributes and social skills. His is a classic case of social promotion, like the ennobling of a vassal or the appointment of a bishop as a cardinal. And it is an example of the rule of reciprocity binding humans and animals: giraffes, traders, monarchs, and rhinoceroses confer nobility and dignity on one another.

For Ganda this was the final step in this chain of courtly exchanges. The king's decision to present him to the pope obeyed the same logic with which the sultan had presented him to the governor, and the latter in turn to his monarch. The circle of gift exchange was closing. Ganda prepared to complete his journey from the periphery or limbo of an imaginary space to Rome, the very heart of the West and of Christianity. Never was this object of exchange, this living creature that had been turned into an object and treated as a consumer good, more human. It was this very mechanism that had inserted him into social networks and conferred on him new roles and a new life—the same social life that Arjun Appadurai attributes to things.[16]

Other disciplines besides anthropology have explored the frontier between human beings and things. One of these is the history of science, with its familiar division of the traffic between subjects who acted and spoke and objects that were made and, paradoxically, spoke for themselves but never said anything. If today it is possible to write the biographies of scientific objects and to claim autonomy and life for other themes and actors (such as X rays or reflector telescopes),[17] it does not seem too eccentric to grant a biography and social life to what was, when all is said and done, a living being, and moreover one who performed social functions that extend down to these very pages.

Ganda was acquired, shipped, and presented, an animal degraded to the status of inert matter. But at the same time he was enslaved and acquired a social function in the courtly life of the happy few. He increased his value, that quality of things that, as Georg Simmel understood, is conferred from outside (by the subjects, obviously, since value is a social convention), and is related to their resistance to our desire to possess them and to what we are prepared to sacrifice to obtain them.[18] So if distance and desire, resistance and sacrifice, produce value, Ganda must have been one of the most valuable objects in the world, worthy of a king, or what is more, worthy of the very vicar of Christ on earth, of the ardent Medici pope who also desires the things of this world.

We know some details of this second Mediterranean voyage. The ship left Lisbon in December 1515, this time laden not only with spices, but also with precious ceramics, gilt pitchers, and silver cups, but none as precious as Ganda.[19] After passing the Strait of Gibraltar, the vessel landed near Marseilles, where our rhinoceros made his last public appearance.[20]

So great was the curiosity aroused by his presence that the French king, François I, and his wife hastened to see him. On a fine day in 1516 they presided over a mock battle using oranges and other fruit instead of munitions. This practice could boast a certain tradition, for we know that some of the festivities at royal weddings included Ethiopian (that is, Negro) boys mounted on elephants who hurled oranges at the public in the streets of Lisbon. It is not clear where the mock battle with Ganda took place; it may have been in the vicinity of Marseilles, probably on one of the small neighboring Frioul islands of Pomègues, Ratonneau, or Château d'If. We do know, however, that the Portuguese captain presented François with a beautiful horse and that the French king returned the favor with the considerable sum of five thousand escudos. And we know what Ganda looked like on the occasion: appareled like a bride, adorned with roses and yellow carnations, and sporting a gilt harness and green velvet fabrics.[21] Beneath this extreme disguise and symbolism enshrined in ancestral and sophisticated rituals of pageantry, foreign to it all and evidently dragged in chains, Ganda was on his way to Rome, where he was to reach the apex of his meteoric social ascent and meet his partner, opponent, and host all in one.

But however exclusive one may be, there is always someone else who has been exclusive earlier or who has been more exclusive. Ganda was not the first pachyderm to arrive at such a dizzying height. In the previous year, 1514, King Manuel I had sent an elephant as a gift to Giovanni di Lorenzo di Medici to celebrate his recent election to the papal throne as Leo X. Giovanni was a worthy son of his father. At the age of thirteen he had been appointed cardinal, thereby gaining access to the Curia. Aged thirty-seven by now, he knew all there was to know about the art of politics and diplomacy. He wanted François I to be made emperor, but ended up supporting Charles V once he had realized that there was no stopping him anyway. Giovanni favored his relatives whenever he could, even creating a duchy for one of his illegitimate sons. The list of his maneuvers would be long.

But Leo X has also gone down in history as a patron of the arts. He was refined, cultivated, and promiscuous. It would be difficult to say whether more or less so than Sultan Muzafar II or Emperor Frederick II, although everything points to his having been in a different league. His devotion to the pleasures of the flesh even led him to have the following motto inscribed on a votive arch at his coronation: "Mars has reigned, Pallas followed, but the reign of Venus is eternal." His pontificate was not confined to venereal pleasures (though never was syphilis as rife among the prelates as then), for he cultivated eating and drinking just as much as sex at his bacchanals under the sign of Dionysus and Aphrodite.

His love of literature led him to appoint scholars and poets as cardinals. As patron of Raphael and Bramante, it was under his papacy that the Basilica of St. Peter was completely reconstructed. He spared no expense. In fact, the extravagance and luxury of his court are often cited as the sparks that triggered the rebellion of Luther, since it was during his papacy that the Reformation began.

The gift of the elephant by Manuel I was part of a diplomatic initiative aimed at obtaining the papal bulls that would seal the line of division between the two Iberian monarchies in the West as established by the Treaty of Tordesillas in 1494.[22] It would not be long either before the Treaty of Zaragoza (1529) would fix the limit of the Portuguese Empire in the East. There could be no better way to claim, confirm, or legitimize Portuguese hegemony in the Orient than to give the pope an elephant, a white one at that, which made it particularly special. Like Ganda, it came from India through the good offices of Afonso de Albuquerque, governor, colonizer, and, as we have seen, dealer in large pachyderms as well as wild cats, for the elephant was accompanied by a fabulous snow leopard (or ounce) and a cheetah, whose elegant and stealthy movements must have contrasted with the ponderous steps of the huge pachyderm.

Like the success of the Mamluks in Florence and of Julius Caesar after the Egyptian campaign, the arrival of the Portuguese diplomatic mission in Rome in March 1514 was a genuinely memorable event. That is how King Manuel I wanted it.

It was the first time that Portugal displayed its exploits of discovery and navigation. The king appointed three ambassadors to the delegation: at their head the prominent navigator Tristão da Cunha (the mission is usually known as the "embassy of Tristão da Cunha"); the diplomat, physician, and humanist Diogo Pacheco, who read a discourse in Latin epigrams; and João de Faria, grand chancellor of Portugal and its ambassador in Castile and Rome. The distinguished mission also included the poet and official chronicler Garcia de Resende, who naturally put the events down in writing.[23]

King Manuel I knew very well the tastes of the recently elected pope and was willing to satisfy him on this unsurpassable occasion with the mission of obedience, as the first visit to a new pope was called in Vatican diplomacy. In fact, the Portuguese monarch took advantage of the opportunity to present his country as the worthy successor to the Roman Empire. As the Portuguese paraded through the Eternal City toward Belvedere Palace, they displayed their parrots, cocks, silks, and jewels, a Persian horse from the King of Ormuz, the wild cats, and the impressive elephant, as white as the pope himself though considerably younger. In fact, the elephant was barely four years old when it entered the Holy See. Some eyewitnesses stated that the snow leopard was carried in a cage mounted on the hindquarters of the Persian horse. The cage was exquisitely finished as a coffer that allowed onlookers to marvel at the treasure that traveled inside it, a natural, live treasure whose value lay in its rarity. The leopard was as beautiful and exclusive as a topaz, for in its natural habitat it only lives above six thousand meters in central Asia. It was a predator of remarkable agility, able to walk on snow and to jump from rock to rock without being heard. All the same, it was easy to domesticate, as the

Portuguese had already seen in some Oriental and African courts.

Apparently Leo X ran enthusiastically through the covered passageway connecting the papal residence with the Castel Sant'Angelo to enjoy a better view of the company. The young elephant, escorted by a Saracen guide and mounted by its Indian mahout, knelt three times before the tower of the Castel Sant'Angelo and even trumpeted before the pope.[24]

Leo X was just as enthusiastic about cavalcades, parades, and the display of exotic animals as his father had been. When resting from the gastronomic feats of his dwarfs, he loved to go and see his Hindu hens, his Persian horses, and his leopards. The white elephant, soon named Hanno, became one of the most famous personalities of his papacy.[25] Hundreds of verses were written about him and we have numerous sketches and drawings of him, although the most famous representation of Hanno, a fresco by no less an artist than Raphael, was irretrievably lost a century later when the architect Carlo Maderno replaced the wall on which it had been painted by the Porta Horaria, thereby depriving posterity of one of the great Renaissance representations of exotic fauna. This at least is the picture that emerges from the preparatory works that have survived and the evidence of those who were fortunate enough to view those scenes in the courtly life of the exiled pachyderm. Raphael also included the elephant in *The Triumph of Bacchus in India*, which was left incomplete at the artist's premature death in 1519. It had been commissioned for the *studiolo* of Alfonso d'Este in his palace in Ferrara, a private area known as the Camerino d'Alabastro, for which Titian also executed several canvases, including his *Bacchus and Ariadne* (1518–1520), which includes exotic fauna but no elephant.

Figure 3. Hanno, Pope Leo X's elephant, attributed to Raphael or Giulio Romano (after a lost drawing by Raphael).

If in some of these versions Bacchus represented Portugal (he was one of its mythological founders, like Ulysses of Lisbon), the elephant Hanno came to be the incarnation of Christianized and domesticated India. The elephant soon became the favorite pet of the pope and of the Roman populace (Figure 3). His custodian was Giovanni Battista Branconio, a privy chamberlain of the pope. Hanno lived in the patios of Belvedere until a proper house could be constructed for him between the Basilica of St. Peter and the Apostolic Palace. He appeared at a reception for Giuliano de Medici, Leo's brother, and at other kinds of festivities. One of these was the large-scale charade organized for the new Petrarca redivivus, Giacomo Baraballo, a rather crazy minor poet whom the pope and his retinue dressed in holy vestments, crowned with laurel, and mounted on the elephant to ride to the Capitoline Hill, the ancient eminence dedicated to Saturn. The cannon fire frightened the beast, and it all ended, predictably, with poor Baraballo crawling in the mud, humiliated.[26] It should be added that the victim was already advanced in age. The humanity of some people consists of treating their peers like animals. When the moment comes, they are all enlisted in the spectacle.

The death of Hanno was preceded by the sinister appearance of Fra Bonaventura, a mystic inspired by Savonarola, who proclaimed himself pope and excommunicated the Medici pope and five of his cardinals. He prophesied the death of them all and also of the elephant Hanno. His twenty thousand followers were not enough to prevent his imprisonment and torture, but it was not long before the five cardinals grew ill and died, one after the other. Even Hanno began to show signs of acute illness. After a detailed scrutiny of urine and blood, the poor creature was obliged to ingest a purgative of five hundred grams of pure gold, but in vain. His death on June 16, 1516, was felt as a curse, a terrible omen. Pope Leo X himself composed an epitaph that the poet Filippo Beroaldo turned into elegant hexameters. This was when Raphael was commissioned to paint the lifesize murals of Hanno (Figure 4).[27]

What was a tragedy for some was the material of farce for others. The poet and dramatist Pietro Aretino, famous for his erotic verses and parodies of courtly customs—the only person spared his wit was Jesus Christ because, he said, he had not

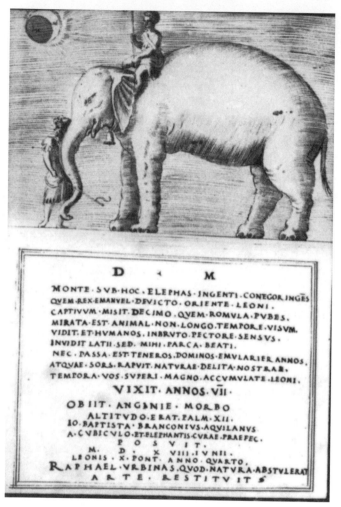

Figure 4. Hanno epitaph, by Francisco de Holanda, sketchbook, 1538 (El Escorial Library, Madrid).

known him—wrote a brilliant satirical poem, the "Testamento dell'elefante" (1516), in which a corrupt cardinal inherited the pachyderm's prodigious genitals.

The whole Vatican court was in crisis. The pope's health, which had never been rosy, deteriorated considerably. His back pains and ulcers, which often left him bent double, were now accompanied by strong depressions, the affliction of the soul that cannot flee from itself, as Montaigne acutely diagnosed. The first stages of the Reform in Germany were beginning to shake the foundations of the Basilica of St. Peter. Raphael, the pope's great artistic protégé, died young in 1519, a decade before the two Iberian monarchies were to nominally divide the oriental hemisphere in Zaragoza with the sanction of papal bulls.

Leo X, our collector of Oriental pachyderms, died a year later. His funeral was a relatively quiet affair: the time of the grand ceremonies and exhibitions, the time of fleshly excess and luxury, had given way to a moment of silence and seclusion. At any rate, five years earlier, when Manuel I decided that Ganda would follow in the footsteps of Hanno, he knew very well what he was undertaking: a diplomatic operation that enjoyed the blessing of the pope and would likely yield fruit. Leo X was prepared to favor Portugal in every possible way. As a token of gratitude for the gift of the elephant, for example, he granted Manuel the highest Vatican distinction, the *Gladius et Pileus* (Papal Sword and Ducal Cap).

When Ganda passed through Marseilles in January 1516, the storm clouds were already gathering, and burst soon afterwards. Before the disasters described above, before the death of Giuliano de Medici, the pope's brother, in March 1516, the first bad omen was the shipwreck of the vessel on which Ganda was traveling. It sank off the Ligurian coast near one of its most beautiful spots, Portovenere, where centuries earlier the Romans had erected a temple to Venus, the goddess whose reign the pope wished eternal.

It is not difficult to see the death of Ganda as foreshadowing imminent events in Rome, but also as a counterpart—suffered by an animal in the Mediterranean—to the Portuguese disasters at sea in distant oceans, which were catalogued much later by Bernardo Gomes de Brito.[28] Ganda did not meet Hanno. Although they followed the same routes—from Goa to Lisbon and from Lisbon to Rome, both doubled the Cape of Good Hope and passed between the Pillars of Hercules from the Indian Ocean to the Mediterranean—their fates were diverse. One was crowned; the other died at sea. Hanno was at the center of the stage of courtly life in Medicean Rome; in a certain sense, the shipwreck of Ganda recalls the tragic destiny of those who dare to cross the oceans.

CHAPTER TWO

Words

The more closely you look at a word, the more distantly it
looks back.

Walter Benjamin, "Karl Kraus"

What did Europeans know about the rhinoceros before
Ganda disembarked in Lisbon? Barely anything. What
was known about Oriental fauna at the time? Not very much,
and most of it legend.

What little was known about the rhinoceros left no room for
doubt. The Portuguese fully expected to see a ferocious animal
built for combat. During his seven months in captivity in
Lisbon, the most memorable event in the life of Ganda, the
only one that no chronicler failed to mention and that no historian can overlook, expresses exactly what was going on in the
minds of his captors, what type of knowledge they had, where
it came from, and what their aims were. This was his famous
duel with the elephant in the Terreiro do Paço.

Let us reconstruct the episode.[1] After Ganda's arrival, it is
unlikely that the Portuguese monarch housed him in the stables of the Paço dos Estaus in the Praça do Rossio. This was
where the other "state pachyderms," as Gomes de Brito called
them, were kept.[2] These were various Indian elephants, certainly

more appreciated than other illustrious visitors, the noble and high-ranking dignitaries who were lodged in the same palace at Estaus.

More exclusive than any of them, Ganda could not be kept there. Doing so would have been rash: elephants and rhinoceroses were natural enemies; their rivalry was inevitable. In fact, this enmity was the most notorious attribute of the rhinoceros, associated with its most striking anatomical feature, the horn made to rip open the belly of its enemy. It should be remembered that to Europeans Ganda was basically the sum total of a horn and armor, made to charge and to resist and to give and to take. His anatomy was designed for combat. He had been brought to Lisbon to pit himself against a rival of his own size.

He would have been kept in the stables of the royal palace known as the Paço da Ribeira, or perhaps in the neighboring Casa da India. Manuel I had abandoned the upper part of the city as a clear indication of where the center of the nation was now situated: beside the river. In front rose the central authority for the organization of overseas trade, the Casa da India, the enlarged version of the old building of the Casa da Mina or Casa da Guiné. That was where the spice monopoly was maintained and where the master nautical chart, the Padrão Real, was produced and updated. The colonial institution par excellence had changed its name at the moment when the Portuguese vessels had moved on to a new continent and ocean.

In the center of these two buildings that played such a notable role in the history of Portugal and of our rhinoceros (Ganda would never have come so far without the king or the empire), and of which hardly anything is left (they were swept away by the earthquake and tidal waves of 1755), lay an exten-

Figure 5. Terreiro do Paço, detail of Lisbon view, in Georg Braun and Franz Hogenberg, *Civitates Orbis Terrarum*, 1582.

sive esplanade or inner patio, the Terreiro do Paço, the forerunner of what is now known as the Praça do Comércio.

At the time, this patio was enclosed by a crenellated wall with barred windows. One of its sides contained a passageway that linked the rooms of the king and the queen, a corridor or gallery decorated with large tapestries that hung down to the ground and functioned as a partition between the corridor and the Terreiro do Paço (Figure 5).[3]

On the morning of June 3, 1515, Holy Trinity Sunday (which comes after Whitsunday), the rhinoceros was taken by Oçém, his Indian handler, to the interior of that passage, where he was kept waiting for some time, in chains and hidden by the tapestries. All the members of the royal family, the principal members of the court, and several distinguished visitors to the city were comfortably seated at the main windows of the palace. Those for whom there was no space there had to find a vantage point from some other window or terrace. The square was filled with an expectant public, all crowded together on the battlements, waiting to attend in communion, as at a religious holy drama, a scene that seemed to have come from the spectacles of ancient Rome.

Although combat between two wild animals, whether of the same species or not, has been staged by many cultures at different latitudes and at different times, during the Renaissance these kinds of games or competitions were intended to reflect the glory of Rome, which is also how they were understood by the crowds. The revived tradition, the *renovatio*, emanated from the splendor of the city of which Petrarch said, when he visited it two centuries earlier, that he was surprised not that it had subjugated the world, but that it had had taken so long to do so. As far as displays and contests between wild animals were concerned, even in Lisbon, Rome was once again the model.

Damião de Góis, the great humanist friend of Erasmus who was portrayed by Albrecht Dürer and who wrote the account from which we are attempting to reconstruct the event, began by referring to Roman precedent. The Romans used to organize duels between two men who had been condemned to death or between two beasts *(alimárias)*.[4] King Manuel was pitting the two strongest and most feared animals on earth against each other.

While Ganda and Oçém waited hidden from sight in the wings, their antagonist, the youngest elephant from the Paço dos Estaus, was plodding along from the nearby Praça do

Rossio, escorted by his Indian mahout. Their passing met with enthusiasm and curiosity from the crowd that packed the streets. When they finally reached the center of the arena, the king gave the order to close the entrance gates and to hoist the hangings in the passageway.

As soon as he saw his opponent in the middle of the arena, Ganda rushed at him in fury, raising a great cloud of dust as he dragged Oçém in his wake until the latter managed to loosen the chains that bound the rhinoceros. The elephant, which was facing the other way and had to turn around quickly, only had time to raise its trunk, trumpet loudly in fear, and endure the attack by its rival.[5]

Both Damião de Góis and Valentim Fernandes, our two eye-witnesses, agree that the elephant fled in terror at the approach of Ganda. Apparently the rhinoceros aimed, predictably, at its belly, and the poor juvenile elephant was unable to defend itself with its short tusks, which were no more than three palms (about two feet) in length. It made for one of the walls at a trot, managed to wrench the wrought iron grill covering one of the windows with his trunk—in spite of the eight-inch bars being "as thick as arms," and did not stop until he had managed to force his entire body through the hole he had made. In the meantime, the proud and victorious Ganda remained in the middle of the arena, in the words of Damião de Góis "very as-sured and conveying to the public by his gestures and move-ments that his victory was unassailable."[6]

These words suggest that our rhinoceros is being treated as a personality endowed with protohuman traits. He is pro-claiming himself the champion of the tournament, a competi-tion prefigured centuries earlier. The duel between the two pachyderms is an entirely social spectacle worthy of the "thick description" with which Clifford Geertz, the father of cultural anthropology, analyzed Balinese cockfighting.[7] The battle between the huge mammals in the Terreiro do Paço is at the same time a profound sport with traits worthy of a Shakespearean drama, staged in a circus redolent of classical Rome.

Obviously the spectacle lends itself to a colonial reading. It manifests the power of the organizer, the monarch whose nation has seized a foreign territory, and exercises symbolic possession by forcing into combat two of its most representative and majestic living creatures. The final destiny of Ganda in this colonial tale would have been to join the white elephant Hanno in Rome in exchange for pontifical bulls and sanctions. The two pachyderms were part of the symbolic exchange connected with the juridical possession of an entire planetary hemisphere. What is their significance in that traffic? It is difficult to gauge with precision, but also difficult to overestimate.

That, however, is not the whole story. The episode raises the question of the relationship of a culture with the past, with inherited forms of understanding about what a living being is, how it is known, what can be expected of it, what its obligations are. The duel encapsulates a whole reading of natural history at the same time that it expresses a more latent conflict that needs to be represented time after time.

THUS WAS SETTLED a combat worthy of the Caesars, a spectacle hailed by half the princes and aristocrats of Europe and about which news circulated rapidly among the major courts and the scholars, humanists, and chroniclers who belonged to them.

In Italy, for example, a Florentine physician from the court of Pope Leo X named Giovanni Giacomo Penni penned and published a poem in honor of the illustrious visitor less than two months after the event. Romans knew about the disembarkation of Ganda, of whom so much was expected. The very possibility of seeing him matched with Hanno was tantalizing. What would happen if they met? What would be the outcome of an encounter between the favorite papal pet and that ferocious beast of the Indies? What would Leo X organize to mark the arrival of the rhinoceros? Probably another duel, and perhaps a kind of reconciliation or a ceremony of submission of the most evangelical kind? No symbolism could have been more powerful. It is uncertain whether the Medici pope would have risked the life of his favorite, but it is equally uncertain whether he would have shrunk from sacrificing any loved creature with the prospect of a magnificent spectacle in view.

We shall never know, but such doubts were already latent in Penni's poem, published in Rome in 1515.[8] These verses reflect a fairly precise knowledge of the arrival of Ganda, enveloped in the fabulous orientalizing rhetoric that accompanied it. Penni refers to the fleet with which Ganda had sailed, laden "to the brim" with pearls, pepper, ginger, myrrh, and sandalwood. He mentions various animals, including monkeys and baboons, and even some unnamed ones that are used to scent the beds of courtesans.[9] And of course our rhinoceros is a treasure more valuable than rubies, the natural enemy of the elephant, a stranger to Europe who walks with shackles around his ankles. Will he come to Rome?

Manuel I had brought him to Lisbon and confronted him with his legend. With hindsight, the scene can be regarded as a full-fledged ethological experiment to see whether things corresponded with words, as they were supposed to do in this primitive world where the court looked up to its king, fire chased the sun (according to Aristotelian physics), and knowledgeable men venerated their ancestors, the great sages of antiquity who continued to guide their paths as masters guide their domestic animals.

In this strange lost world where the word, the emanation of authority and religion, still kept its therapeutic and liberating force to cure and to save, where the word had not yet been reproduced endlessly, its meaning weakened by repetition before being eclipsed by the emergence of the image and visual culture, observable phenomena were expected to behave as words dictated. Natural facts had to pay tribute to words.

Renaissance natural history, like other disciplines related to scientific activity, was a body of knowledge based on the importance of the word, whether written or spoken by authoritative voices, starting with the word of God. But the same was true of other words with power,[10] those that the great sages had written in texts since antiquity, those books that were copied time and again over the centuries: books that were translated, annotated, abbreviated, condensed, their words repeated, overlapping, confused.

Half of Renaissance science consisted of an exegesis and a gloss of the words of antiquity; the other half comprised a dissection of the world that nevertheless claimed and accepted the authority of the ancients. As Anthony Grafton has skillfully demonstrated, they were all defenders of the text, ancient and modern, even though hindsight has tended to concentrate on certain programmatic declarations of the moderns.[11] Questions of philology were of crucial importance in this context. Nothing mattered more than to decode exactly what Aristotle had really said, why Averroës distorted Aristotle's ideas, at what point Avicenna blurred the outlines of genuine Galenic theory, and which books were misrepresented as genuine copies of ancient texts and which reflected the exact truths of the originals. Taken to extremes, these endeavors raised questions about the true words and names of things and of sentient beings before Babylon and Babel introduced confusion and plunged humanity into sin and ignorance.[12]

The insistence during the first stage of global exploration on fitting the newly discovered lands in with the sacred scriptures and with Ptolemaic geography is a revealing example of how knowledge functions in general terms and is undoubtedly a good correlate of how this persistent epistemological principle operated. When Lorenzo Valla said that his predecessors had excelled in every kind of research, he was not thinking solely of the Latin language, or even of *literae humaniores*.[13]

A rhinoceros had to be what others who were more learned had declared that it was. The influence of the written word and the authority of the ancients, combined with an awareness that knowledge consisted of a restoration of truths that had been eroded by time—an authentic return—also weighed on our armored pachyderm; or rather, on the contrary, he weighed on them: knowledge and authority were more weighty still and exercised a stronger force of attraction, because in the premodern world the center and the heart of knowledge lay in the word.

So as soon as Ganda arrived in Lisbon, Manuel I confronted him with his legend, put him in direct contact with how the word had said he ought to act. This natural history was deontological, verbal, finalist, and teleological. Like the four elements and everything else in creation, the rhinoceros had its obligations and its natural place. Pitting it against an elephant was no more than an experiment before experimental philosophy—in other words, a scholastic experiment. According to the paradoxical inversion of values in this game, the objects that were manipulated were intended to confirm and obey the written word instead of skeptically testing it. This is why Damião de Góis could write that the king wanted "to see with his own eyes *(por experiencia)* what the ancient writers had written regarding the natural enmity between elephants and rhinoceroses."[14] The courtyard was closed off hermetically like an isolated and controlled laboratory. The purpose of the test, however, was opposite that of the modern notion of an experiment. Instead of observing events in order to disrupt or at least question the stability of a prescribed world, the objective was to underline the nature of things and to confirm the solidity of what we know about them.

Little had been written about the rhinoceros, but those few words were widely accredited. The legendary hostility to the elephant came from two main sources—the *Natural History* of Pliny and the *Geography* of Strabo—plus later compilations like that of Gaius Julius Solinus and the information provided by Diodorus Siculus and Isidore of Seville. The *Natural History* by Pliny (ca. 23–79 AD) was the main reference work of natural history for centuries; it was printed for the first time in 1469 and was reissued for more than ten editions in Italy alone before 1500. In this work, the author's discussion of the rhinoceros is subordinate, an appendage to his remarks on the elephant; it features in a review of several Oriental species written soon after the appearance of the giraffe at the games organized by the dictator Caesar.[15]

In the strongly hierarchical zoology of the ancient world, the elephant occupied the principal place for being "the largest and the closest to human sensibility." Following Aristotle, Pliny confirmed that the elephant not only understood the language of its country, but also was able to memorize tasks, enjoyed love and glory, and even worshipped the stars and venerated the sun and the moon. Its religious observances, its goodness and intelligence, its social and moral virtues made it a blessed animal.[16]

The rhinoceros was attributed with the opposite characteristics. Pliny mentions one with a single horn that had been seen in Rome in the celebrated games held in honor of the victorious return of Pompey: "This is the second natural enemy of the elephant. It gets ready for battle by filing its horns on rocks, and in the encounter goes specially for the belly, which it knows to be softer."[17] Indeed, while the skin on the back of the elephant is very hard, the skin of its belly is soft. Pliny took this information from Strabo, and it was later repeated by Solinus, Isidore of Seville, and many others. The elephant's belly was its Achilles' heel, the spot that its second enemy, the malevolent rhinoceros, always sought.

Who was the elephant's first enemy? The *draco*, an imprecise term in Pliny's usage, which may refer to serpents and by extension to reptiles.[18] The rhinoceros thus emerged as a marginal and secondary enemy of the elephant, second to the *draco*.[19]

The other main source is Strabo (63 BC–23 AD), the Greek historian, geographer, and traveler whose death more or less coincided with the birth of Pliny, for whom the Greek author was required reading. Strabo was the first to describe and name the rhinoceros in a passage of his *Geography*, perhaps the best description of the world in antiquity, an extensive treatise that was first printed in Rome in 1469 and in Venice three years later (in Latin and by the same printer who had published Pliny's

Natural History). Among other things, Strabo's work is perhaps one of the most comprehensive works on Asia ever written by an inhabitant of the Mediterranean.[20] The name that he chose has persisted down to the present: ρινόκερως, rhinoceros, literally "nose + horn." Apparently Strabo saw one with his own eyes in Alexandria, although he also drew on secondhand accounts. He alluded to its already proverbial hostility toward the elephant and gave a fairly precise physical description: it was smaller than the elephant, similar in color, but perhaps more like a bull in size; it resembled a wild boar, particularly the forehead, although its single horn rose directly from the nose and was harder than any other animal's. The rhinoceros used its nasal horn as a weapon in the way that a wild boar uses its tusks. The other feature of its anatomy worthy of mention was its hide, with powerful welts encircling the body from the head to the belly like the collars of serpents.[21] At first sight, then and now, Ganda or any other member of his species was the sum of his two most visible attributes: a horn and body armor.

Although it could also be found westward of the Ethiopians, Strabo claimed that the rhinoceros lived not far from the habitat of the remarkable giraffe *(camelopardalis)*, Cynocefali (improbable dog-headed humans), and other extraordinary beings beyond the Bab-el-Mandeb Strait, or Gateway of Tears, that separates Somalia from Yemen—in other words, Africa from Asia.

Pliny and Strabo were the two main sources on which the fame of Ganda rested when he disembarked in Lisbon. Renaissance natural history was a bookish science founded on established authority. If Strabo was the original source for the first descriptions of the rhinoceros and the greatest authority on questions bearing on Asia, Pliny the Elder was the main reference source from the classical world when it came to natural history. Coming behind the Bible but immediately after Aristotle—the Stagirite always headed the list of all the sages on all forms of knowledge, from logic to zoology—Pliny was prominent in European science until the eighteenth century for a variety of reasons. On the one hand, his vast encyclopedic work (only a small part of which this Roman military commander, whose introduction to zoology came from his love of horses, actually wrote) had the virtue of bringing together what many others had written about fauna, flora, ethnography, geography, therapeutics, and medicine; the thirty-seven books of the *Natural History* were the storehouse of a large part of the classical tradition. The rhinoceros, about which Pliny paraphrased what Strabo had written, is a case in point. A library could do without other books, but not without this one.

Pliny also found the right intellectual space in Renaissance humanism. Or was it the other way around? Were the humanists eager to detect in Pliny a model for the sort of knowledge they wanted to present anyway? An allegorical natural history, attentive to the moral aspects and exemplary dimensions of the accounts of each species, as well as their practical uses and the ailments that each plant cured; a universal natural history with room for the exotic charm of the customs and beliefs of the peoples on the margins of the known world; a sympathy and antipathy, attraction or repulsion, between the cosmos and humankind, between the elements or humors—between opposite pairs like the elephant and the rhinoceros; to sum up, the whole of this disciplinary format and style of producing and elaborating knowledge harmonized astonishingly well with the aesthetic and philosophical horizon of Renaissance humanism.[22]

On top of all that, Pliny was an elegant stylist. Attuned as he was to the beauty of the world, maddeningly erudite but simultaneously entertaining, his prose made and still makes reading him a pleasure. In comparison with Aristotelian logic and its desire to reduce knowledge to a few general principles—a reductive characterization of Aristotle's work but one that defines his aim—Plinian science breathed the delicate air of what is singular and unrepeatable. Although Aristotle's *History of Animals* is the main source for Pliny's *Natural History*, the latter is less severe, more proximate, and more artistic. Lyrical rather than taxonomical, Pliny is always likable and rewarding. There is little doctrinal about it; on the contrary, Pliny guides the reader through the nooks and crannies of nature with interesting, astonishing, or simply marvelous stories: of mares who are impregnated by the wind, the healing properties of a viper burnt in a vessel of salt, the prognosticatory value of swarms of bees. His repertoire is endless. Neither excessively credulous nor overly skeptical, he takes an interest in everything. There is nothing that does not arouse his attention, and it is precisely this virtue of curiosity that was being celebrated during the humanist Renaissance, the great passion of the soul that preceded and was a precondition for knowledge (an idea also formulated by Aristotle, whose ascendancy, as we see, is difficult to shake off, no matter what the topic may be).[23]

Plinian science linked up admirably with the interests of the humanists. Pliny's love of Rome, his timeless knowledge, and the combination of his Stoic gaze with his enjoyment of life in its different manifestations made him a classic—one of the very classics that the humanists were to revive and emulate. Renaissance natural history, which bore its medieval past of bestiaries and stone carvings, allegorical readings of the lives of animals as exemplars for the lives of humans, found in Pliny an inexhaustible treasure house. He was the model for a natural history with a moral horizon.

There was yet another aspect of the work of Pliny that especially endeared him to Renaissance naturalists and scholars: his passion for the extraordinary, his relish in compiling the teratological diversity of creation. This set him apart from Aristotle, who was more interested in regularities and laws. It was also something that would make Pliny the primary point of reference for the natural history of the New World of Asia and America, from which the first of a flow of extraordinary phenomena, monsters, and natural prodigies were already arriving.[24] It is easy to understand why one of the channels by which Pliny survived in the Middle Ages was Solinus's potpourri of marvels, which included a large number of the ancient descriptions of Oriental fauna, including the rhinoceros, and which often circulated under a title containing another concept associated with what is miraculous or prodigious: the memorable.[25]

This explains why many in the sixteenth century liked to regard the naturalist Ulisse Aldrovandi (1522–1605) or the physician Francisco Hernández (1514–1578) as the new Pliny, as men who continued the classical author's work. Many indeed sought this title for themselves even if few deserved it. If the aim of natural history had been to record memorable facts, there could be nothing more memorable, more in need of being recorded and collected, than the unfamiliar and unprecedented. And what was more capable of overturning accepted ideas and provoking astonishment than the fauna of Asia or America? They were the most marvelous and strange living species in the eyes of the Europeans (each in its own way: the fauna of America for being entirely new and unexpected, the fauna of Asia for being enveloped in a farrago of snippets of information, distorted or legendary stories). Without doubt, Pliny's form of natural history was the ideal discipline in which to unfold the language of marvels, the penchant for the extraordinary, a key feature of modern science being the science of the age of European expansion and the development of the colonial enterprise.[26]

Finally, there was an imponderable element in Pliny's conception of nature that would resonate within Neoplatonism and that interests us particularly here. Many passages show that Pliny connected external natural forms with what everyone bears inside, whether in the mind, the soul, or any other region that is more or less material, spiritual, or at any rate difficult to locate. With its mixture of centaurs preserved in honey and accurate empirical observations, basilisks, and other striking representatives of zoological fantasy with a whole catalogue of very precise annotations on the morphology, uses, and customs of animals—information that indicates a remarkable scientific acumen—Pliny's work can be read as a call to pay attention to the importance of what Italo Calvino, following a time-hallowed tradition, called "the alphabet of dreams," the presence of the world of dreams in the lives of human beings and in their forms of knowledge.[27] Pliny's work, in short, throws us back to the role of the imagination, whose mysterious workings are so central to this history.

WE IMAGINE WHAT we surmise but cannot see, what takes form in the mind without being preceded or confirmed by the experience of the senses. We also imagine what does not exist but should. Moreover, what we prefigure or dream tends to be connected with desires and fears. In this sense there are few territories that are more oneiric for Europeans than the Orient, an almost fictitious space whose nature for centuries bore the weight of the social and moral projections of the desires and fears of the West.[28] Let us examine our pachyderm from this perspective, a perspective that incorporates a certain symbolic and moral reading of the fauna of the East.

The history of the prefigurations of the Orient goes back to at least the fifth century BC, when a Greek physician was taken prisoner by the Persians (the East and the West are always taking each other prisoner, or at least trying to). The prisoner, Ctesias of Cnidus, who enjoyed the favor of Artaxerxes, created a niche for himself in the Persian court and managed to piece together an image of India from the stories that he heard from diplomats, merchants, and other traders in the embassies without having set foot there himself. He had to imagine it. Ctesias wrote that in India there were springs filled with liquid gold, dog-headed people, and prodigious animals like the manticore, the voracious man-eater with a human head, lion's body, and reptile's tail.[29]

The sun in the Orient must be ten times larger than in the Mediterranean, causing a suffocating heat and the generation of exuberant, oversized flora and fauna. From then on, the Orient would be unrivaled as a privileged place for hybrids, teratology, and other imagined natural phenomena. In later centuries Europeans claimed to have seen human beings in Asia with their heads in their breasts, eyeless, as well as androgynous and quadrupedal people.

Ethnography was not the only fantastic discipline. Ctesias also described a strange wild ass with a white mane, blue eyes, and a sharp horn on its forehead. It was the unicorn of the kingdoms of Hindustan, as Pliny later confirmed. Its body was like that of a horse, its head like that of a stag, its feet like those of an elephant, and its tail like that of a wild boar. All of this was crowned, of course, by the long black horn in the middle of its forehead. The unicorn and the rhinoceros: although Pliny distinguished them, it is certain that from then on, descriptions by others frequently borrowed from, confused, and merged them. These are the collisions that take place over generations of intellectual traffic.[30]

Myths about unicornlike creatures are exceptionally rich and widespread. They can be found in very diverse cultural traditions. For the ancient Chinese, for example, the *qilin*, an animal much like a unicorn, occupied a privileged position alongside the tortoise and creatures comparable to the Western dragon and phoenix, all four being endowed with high spiritual value. In Persian culture the unicorn is connected with danger and cruelty. It also featured in Greek cosmogony giving suck to Zeus: perhaps its horn was the cornucopia.

But its weight in the Western tradition is due to its supposed presence in the Old Testament, certainly the result of a poor translation—another of the creative risks of commerce and cultural transfer. It is more likely that the few times it is mentioned in Psalms, Numbers, Deuteronomy, and Job the reference is to a buffalo. This, however, did not stop the unicorn from playing a major role in the Christian symbolism of the Middle Ages: in its submission to the virgin it resembled Christ in his submission to the Virgin Mary and expressed the neutralization of sensuality by innocence.

Its polysemic nature, in fact the ambiguity of its connotations, has been proverbial. Carl Jung went into detail about the relationship between the unicorn and alchemy: Was it an instrument of the devil or of salvation, voluptuous or spotless, brutal or pacific?[31] East and West, blessed and cursed—why are we human beings so keen on binary structures? At any rate, what is certain is that it was the seminal and therapeutic properties of its horn—which was clearly its distinctive attribute and the object of so many conjectures—that rendered it a savage, miraculous, and masculine animal, a mercurial creature.

The words "À mon seul désir" displayed on one of the famous six tapestries known as *The Lady and the Unicorn* from the abbey of Cluny, which were woven in Flanders shortly before Ganda's arrival in Lisbon, appear precisely in the context of a unicorn associated with virginity, but also with the senses and with love (Figure 6).[32]

Without doubt there were similarities between a rhinoceros and a unicorn. The most visible was the miraculous, healing, or threatening horn, an appendage that invariably signified potency and fertility, be they royal, divine, or diabolical. A horned deity, for example, was frequently to be found in some pagan religions of central Asia. But Christianity took it upon itself to demonize the horns, just as the West has demonized and eroticized much else that is strange and remote. The second major similarity is that both animals come from the Orient, a poetic, mysterious place of promise and danger.

A text appeared in third-century Alexandria that would pervade the orientalist imagination in later centuries. *The Life and Exploits of Alexander of Macedonia* inaugurated the cycle of versions of the legendary travels of the great conqueror.[33] India was already being consolidated as the land of marvels par excellence. After Alexander's interview with the gymnosophists on the banks of the Ganges—wisdom and poverty are just as much a part of the stock in trade of our vision of India as

Figure 6. Detail of the tapestry cycle, *The Lady and the Unicorn* (*La Dame à la licorne*, Cluny Abbey, Paris).

eroticism—the Macedonians were attacked by two-headed serpents, horned vipers, crabs with impenetrable shields, and giant bats. This array of fabulous fauna would have been incomplete without a monster larger than an elephant, completely black, with three horns on its forehead, a kind of deadly rhinoceros. Later on, Alexander reached Nisa, the city where erotic processions and bacchanals were held in honor of Shiva, whom the Greeks identified with Dionysus.[34]

In the course of the Middle Ages various versions of these and similar events continued to shape the image of India. Instead of Dionysus, the Christians tried to find traces of Saint Thomas there. Little further evidence on the rhinoceros was forthcoming, however, though this did not stop the bestiaries from continuing to include this extraordinary animal and to confuse it with the unicorn. The technique of hunting unicorns using the ruse of the scent of virgins to put them to sleep and thereby enable them to be caught is sometimes applied to the rhinoceros, a wild animal that could only be subjugated and tamed by divine grace and intervention.

In the twelfth century the rhinoceros served Abelard as an allegory of the conversion of Paul, while our argument is strengthened by the fact that, in a poem on divine wisdom, Alexander Neckham managed to write that the dragon (symbol of sin) and the rhinoceros (symbol of ferocity) were allied against the elephant (the incarnation of good). One of the most spectacular cartographic documents of all time, the Hereford Mappa Mundi of about 1300, located the rhinoceros in Egypt, near the phoenix and the scorpion, although—as the legend states—it came from India like the manticore and the unicorn, symbol of the strength and kindness of Christ the savior.

Few advanced so far in the terrain of orientalism as Marco
Polo (1254–1324).[35] The Venetian tried to put things in their
place, another common habit among merchants and one that
usually means uprooting them and putting them somewhere
else. He did so with diverse results. First, he consolidated certain
topoi that had been in the making for some time. With regard
to India, for example, he asserted that on its east coast, "the
richest and most splendid province in the world," the inhabitants
spent the whole year in a state of total undress. There on the
Malabar Coast the sea was so rich in pearls that the king was
covered in them from head to foot. He wore nothing but these
pearls when he appeared to one of his five hundred wives, the
same number of wives attributed to Modofar, the sultan of
Cambay, who had presented Ganda as a gift. The account probably repeats a topos rather than indicating a royal privilege, but in
any case it specifies a number that no one else could rival, even
though they all practiced polygamy. Here again we find Eastern
riches and voluptuousness.[36]

Although Marco Polo is known as a scrupulous and objective observer, he presented his material in a marvelous light, as
one might expect from a merchant: the wealth, treasures, and
precious stones; the boundless luxury of the Asian courts;
and of course the palace of the Great Khan, the architecture
dreamed up by the emperor that ended up inspiring Samuel
Taylor Coleridge, with the symmetry that Jorge Luis Borges
wanted to see between centuries and continents and that Italo
Calvino revealed in *Invisible Cities*.

The Venetian bore witness to the existence of dog-headed
people on the island of Andaman. He collected and spiced up
legends and fabulous stories: Prester John, the cultural legacy
of Alexander the Great in the steppes of the Byzantine Empire,

the miracles of the Nestorian Christians, the dry tree, the red apple. These legends were intended to justify the superiority of Christianity, its survival in Asia, germ of its hypothetical restoration. Asia had to be converted: the theme was conversion, a variant of domestication.[37]

Marco Polo does not seem to have confused the rhinoceros with the unicorn. He does describe an animal with the hide of a buffalo, the head of a boar, and a single black horn, but is quite emphatic that "they are not at all such as we describe them when we relate that they let themselves be captured by virgins."[38] The animals were even more clearly distinguished by the Dominican Jordan de Severac, who traveled in Asia and wrote *Mirabilia* in the fifteenth century. His contemporary Sir John Mandeville, such a fascinating person that we cannot even be sure he ever lived, crammed his popular book on the marvels of the world with all kinds of tall stories about Asia. He claimed to have found the remains of Saint Thomas in India, as well as men whose lives were spent in dissipation, and he possessed a considerable repertoire of fantastic zoology, but he never added anything about our real or imaginary single-horned creature.[39]

By then, however, Marco Polo had already made something marvelous of reality. He had written about burning stones (anthracite), paper currency, paved streets. These were extraordinary things for Europeans, unheard of and therefore incredible. Although a large part of the account's poetic license and humanism is due to the intervention of Rusticello of Pisa, his fellow prisoner and coauthor, Marco Polo's own vision was that of a merchant, and his work was that of every traveler: to trade in things, words, and legends, to move them from one place to another: from the Book of Daniel to Tartary, from Cathay to

Venice, from the Tigris to the Volga (and later to the Orinoco when Columbus would end up inaugurating a trade as successful as it was surprising, the transport of the Orient to the New World). From the unicorn to the rhinoceros, from books to the world, and vice versa—it was a never-ending transfer process, the conversion and domestication of every soul and every region. But it was also a transfer of the potent lure that the Orient never failed to hold for the West in the form of promises, possibilities, and other (re)naissances or (re)births.

GANDA'S HORN THUS became the site for the projection of a whole series of expectations, fables, and magical properties associated indiscriminately with the unicorn, sexuality, wildness, and the Oriental world. For Europeans, the rhinoceros had for centuries been more invisible than their own unicorn. Once the Portuguese had found it in India, everything proceeded as one would expect. True, it matched all the expectations of Oriental fauna: it was enormous and chimerical, made up of fragments or parts of different animals. The result: a gigantic and weird, lascivious, and violent Ganda in need of capture and domestication. And it had to be pitted against an elephant to confirm the wisdom of the ancients and to reenact for the umpteenth time the eternal duel between good and evil (Figure 7).

The elephant was the first of the big mammals or terrestrial animals in this heavily moralizing and allegorical natural history. Aristotle and Pliny had said so, and everyone else followed them. The elephant was the first because by virtue of its faculties—rationality, benevolence and even religion—it was close to humanity. It is significant that in his account of the

Figure 7. The fight between the rhinoceros and the elephant, in Ambroise Paré, *Discourse de la mumie, de la licorne, des venins et de la peste*, 1582.

duel in the Terreiro do Paço, Damião de Góis spends so much time explaining who the combatants were. The rhinoceros occupies a semiclandestine, sinister place. Much less is known and said about him than about the elephant, the undisputed favorite among the animals, the most sensitive and the most prudent.

In antiquity, Góis tells us, collecting information from the three or four sources mentioned above, there were elephants who could read the Greek alphabet and even one who learned to write with his trunk. In the Indian kingdom of Narsinga, an elephant had even managed to speak to ask for food, he adds, uniting the remoteness of space with the remoteness of time and bringing the item up to date—in other words, universalizing it and confirming an absolutely fantastical statement.[40]

The compilation of Solinus declared that elephants were so attached to their homeland that it was advisable when embarking them for transportation elsewhere to promise them that they would return. Góis goes even further. When Manuel I sent the future Hanno to Leo X, the elephant had become so enamored of Lisbon that the king was obliged to use his Indian keeper as a translator—a genuine mediator between the language of sovereigns and that of pachyderms—to persuade him to go. The monarch's message was that he was sending the elephant to an even greater lord who would treat him better. If the elephant did not feel at home with the pope in Rome, Manuel promised to bring him back. Upon hearing these words in translation, Góis reports, the elephant wept with emotion.[41]

The chronicler cites Pliny on the benevolence of elephants: if people get lost in the forests, the elephants guide them back to civilization (unlike people, he could have added, who remove the elephants from the forests without ever returning them). The elephants act as guides: there is here a reciprocal relation between subjects who exchange roles depending on the situation. They enjoy a semihuman status, as is confirmed by the example Góis recounts of two elephants sent by King Antiochus to ford a river; he writes of the timidity of the one called Ajax and the boldness of the elephant Patroclus, who was rewarded with the rank of captain and all its trappings. The King of Cochin, the port where Ganda was embarked, had a diligent elephant in his service that used to go shopping for its own food with its wages.[42]

The elephants were the personification rather than the incarnation of strength and prudence. Who were their natural enemies? Góis states that nature had created other creatures that were at permanent war with them. He recalls what Diodorus Siculus had written on the wise serpents (cobras) that lay in ambush in the swampy and humid places where the elephants stopped to drink. The serpents would bite them in the feet and the eyes to immobilize and blind them and suck so much of their blood that the elephants collapsed on top of them, thereby spelling the demise of both. When this happened, the blood that flowed from the scaly skin of the serpents (*dragão*) was of a color very like human blood.[43]

The other bitter enemy of the elephant was the rhinoceros, which the people of India called Ganda. Our chronicler identifies it as the beast *re'em* mentioned in the Old Testament books of Numbers and Job and successively translated as buffalo, bison, unicorn (as in the King James version: "God brought them out of Egypt; he hath as it were the strength of a unicorn" and "Canst thou bind the unicorn with his band in the furrow?"), and even rhinoceros.[44]

Doubts about his character emerge. All the classics had emphasized that he had the strength of an elephant, but less noble intentions. The rhinoceros is regarded as a creature whose nature is opposed to the elephant's. When confronted with the goodness, sense of justice, and sensibility of the elephant, Ganda perforce occupies the place of unfettered ferocity and brutality. He does not look to the heavens or the stars, or worship the sun and moon as the elephant does. Góis always portrays him with his head pointing down to the ground, raising a cloud of dust: "They live almost like pigs, throwing themselves into mud and pools. They move with the head so low that they almost seem to be dragging their muzzle through the soil."[45]

His horn was sharp and as hard as iron. He honed it on a stone before aiming at the elephant's belly. And yet it was the source of an antidote, according to a widely held belief. The curative properties of rhinoceros horn, which resemble and are confused with those of the unicorn, are mentioned by ancient sources and are repeated time and again in the Middle Ages. In some royal courts the horn in powdered form was used to detect poisoned food or drink. It was the miraculous medicinal appendix of an evil-minded creature. Belief in its qualities as an aphrodisiac has been practically universal in China and the Mediterranean over thousands of years down to the present.

Ganda's horn is a phallic and very striking protuberance, both for its composition (unlike most animal bone, it is made of keratene) and for its pointed contour. Paradoxically, the very object that has always been considered an attribute of virility and fertility is the reason why the species is on the verge of extinction. It is not at all clear whether it will bury its spike in its opponent's belly or do something else in its combat with the personification of good, the elephant. Good needs to be fertilized by evil.

Perhaps the duel in the Terreiro do Paço was less a ritual to domesticate the savage and control the pleasures of the flesh than a celebration of his power. Ganda's opponent was the very youngest of the elephants available,[46] and this was not a casual choice. The organizers of the duel, with the king at their head, wanted to witness the victory of this ancient, remote, dangerous, carnal, and occult evil. The duel was rigged. To confront the rhinoceros with a baby elephant guaranteed an unequal battle. It was an act of cruelty that shows how much the triumph of evil is desired and how unscrupulous we can be to ensure that triumph.

The same is true of Damião de Góis, whose account betrays a fascination and perverse astonishment produced by the victory of the rhinoceros, which is nothing but the triumph of a primal, ancestral, animal vitality. The rhinoceros drags itself along, head down, and comes from a remote corner of the world. He has barely emerged from the world of dreams and shadows to occupy center stage in this oceanic observatory formed by Lisbon and its square. The unicorn and the dragon were also creatures that lurked hidden in caves or swampy, gloomy, pestilential places. What the fleet has brought from Goa to deposit at the feet of the king and even of the pope is the *Oriente mirabilius* of Isidore of Seville, the oneiric horizon discussed by Jacques Le Goff, a natural chimera in the flesh.[47]

Is the aim to capture and tame this beast, to subjugate it completely? To see it in action is at the forefront of the spectators' minds. To understand the strange tournament of the Terreiro do Paço we should venture to press against or skirt the limits of the real world, perhaps penetrating the realm of the unconscious.

The duel between good and evil has to be staged. The combat between the two big pachyderms represents the struggle between sensuality and rationality, between carnal and spiritual love, a sort of raging fury of Yahweh against the New Testament Christ the Savior. Ganda has a lot in common with prehistoric animals and those of the Old Testament. He bears with him a legend that torments him and prescribes him as a necessary, primitive animal.

The Europeans did the same with him as the Aztecs did with Cortés at the time, and as the Hawaiians were later to do

with James Cook: they integrated him with their body of beliefs, installed him in the midst of a ritual to represent the role
previously assigned to him. Ganda did what was expected of
him. He did so with such success that he was exalted to the
highest position, to that central point in the Mediterranean
where the past always surfaces and where the actions of em-
perors and gods meet. But before reaching the central point of
his journey, before completing his encounter with the elephant
Hanno, with Pliny, with Rome, with the church, with redemp-
tion, with whatever lay in store for him in that other court of
miracles,[48] our rhinoceros was irretrievably drowned at sea near
Portovenere, the port of Venus, goddess of love, "the oldest, the
most perfect, the wisest," as Ficino called her.[49]

Ganda suffered shipwreck and was buried by the waves after
having been put in chains and dragged to the Mediterranean
by that torrent of "words, words, words" that furnished Hamlet
with a reply to Polonius's question "What do you read, my
lord?," that help princes and philosophers to pass the time, that
hold death at bay before what remains, silence, takes over.

CHAPTER THREE

Print

Art does not reproduce the visible; rather, it makes visible.

Paul Klee, *Creative Confession*

There were no survivors. From the captain, João de Pina, down to the last of the men, the crew lay on the ocean floor or floated weightless near the Ligurian coast. Ganda must have been trapped in the hold when the ship went down. Now he joined the crew in death, another corpse in the Mediterranean.

Witnesses of the shipwreck suggest that even in death he remained a precious object, a sunken treasure. They do not mention a single attempt to rescue anybody or anything from the water except Ganda. Some claim he was still alive when eventually he was beached, others that he was dead and that they skinned him then and there on the sands to send the hide to Leo X. Some assert that a cast was taken, others that Ganda was dissected and sent to Rome. One of our sources is the great humanist and bishop Paolo Giovio, among the most highly reputed chroniclers of the century and a major source on the Italian Wars (1494–1559), but also a physician, professor of natural philosophy, and even collector of *naturalia* and *artificialia*. He gives his version of the story a dramatic twist: If

Ganda failed to survive, he says, it was because the coast was precipitous and because he was in chains—a detail that recalls his condition of subjugated prey, his almost slave status, somewhere between humanity and commodity. Once he was beached, he was embalmed so that, as Paolo Giovio declares, at least "his true image and size" could arrive in Rome.[1]

Is that really how things went? It is difficult to know, but in any case it expresses the powerful desire to see him and to touch him, to store him as a treasure, whether dead or alive. Ganda was the Oriental prodigy of the moment. His hide was a trophy. But we all know that was not how he secured fame and immortality.

Instead, he owes his fame to a printer: Albrecht Dürer. The history of Dürer's engraving of Ganda embraces an entire compendium on the relations between science and art, a repertory of questions that speaks to us of the birth of the modern world and of how we manage to know and see what we know and see. It goes without saying that, without the famous woodcut, Ganda would never have reached the twenty-first century or been the topic of these pages.

If the duel in the Terreiro do Paço linked him with the legendary past, the engraving would project him into the future. If words tied him to the ancient world, to classical, allegorical, and moral natural history, the image would draw him into the present. Thanks to art and technology, to the dominance of the emergent visual culture, Ganda was rescued from the seabed, a survivor of his captivity, of his own shipwreck, misfortune, and oblivion, the decomposition of his flesh beneath the waves and its sea change into plankton. He was transformed into something that, even if not eternal, seemed to be so, into something that might not guarantee immortality, but did guarantee universality: his copy. For the ability of art and technology to reproduce and disseminate an image across time and space is a power akin to granting immortality: "So long as men can breathe or eyes can see / So long lives this, and this gives life to thee."[2] Ganda was transformed into Dürer's rhinoceros.

THE HISTORY OF the composition of a drawing, and soon afterwards a woodcut of the rhinoceros, is connected with two characteristic aspects of the German master: his interest in the representation of natural forms and his exploration of the techniques and trade of the engraving.

As far as the former is concerned, it should be recalled that Albrecht Dürer (1471–1528) was one of the outstanding artists of all time in the field of zoological and botanical illustration, as well as one of the most concerned about verisimilitude. Among his contemporaries, Leonardo alone can be considered his equal. The two men shared a common interest in the anatomy of living beings, the physiognomy of a landscape or a human face, and other themes linked to the morphology of life. They shared the vocation to know, discover, and experiment with nature itself. Like Leonardo, Dürer was a great scientist, a theoretician of his profession, and an untiring investigator of natural forms. Also like Leonardo, Dürer knew how to combine a great empirical, one might say technical, decidedly artisanal dedication with the epistemological concerns of a man of science in the "theoretical" meaning of the word. He had an unusual capacity to ask himself questions, to learn new things (he had to study mathematics and many other subjects throughout his life), and a propensity to ask philosophical and epistemological questions.[3]

As Erwin Panofsky has shown, for Dürer art without science (without *Kunst*) combined imitation without reflection with mere conventional practice; it was shackled to tradition and to practical skill or common usage *(Brauch)*.[4] In fact, the theoretical and pedagogical work with which he culminated his career—his three treatises on geometry and perspective, fortification or mixed mathematics, and the proportions of the human body—can be considered one of the most complete and advanced intellectual achievements ever by a painter. In Germany his supremacy in this field was absolute. We have to go to Italy and figures like Alberti and Leonardo in order to find such a striking scientific talent, such a genuine interest in the way things are and how we can find out about them.

Dürer elevated and dignified the condition of the artist. Permanently surrounded by humanists and intellectuals (which caused him major difficulties in his domestic life), he pushed to make art more responsive to the new worlds of philosophy and science. Or perhaps it was the other way round, that philosophy and science threw themselves into the embrace of the arts, especially painting and drawing, to seal the profitable alliance that emerged in the modern world between iconography and truth, between the visualization and the knowledge of a fact.[5]

Let us think of the importance of sensory experience in the context of modern science, at least in that part of modern science based on empiricism and direct contact with the natural phenomena under study. And let us think of the primacy of vision above the other senses. The rise of experimentalism, the importance of the disciplines associated with observation (astronomy, anatomy), and the use of instruments that extended or perfected vision (the telescope, the microscope) are just some of the factors that express a very generalized phenomenon: the superiority claimed for images over words, of tangible and tactile experience above the authority of the ancients, and of the senses above reason.

This elevation of sight over hearing and the other senses reflected in part the revival of the old Platonic analogy between the eyes and the soul as the core question of human understanding. In Platonic philosophy, of course, the world of ideas was superior to and sustained the world of appearances: sight could only imitate, duplicate, follow, or record what the spirit was doing or had done. But this did not alter its primacy among the senses. If there was a faculty or capacity linked with what was most elevated and necessary to knowledge, it was sight. The eyes were the windows of the soul.

The connection between sight and understanding is an ancient idea, found in both Platonism and Stoicism. Various passages in the Bible associate the word with faith and sight with knowledge. When Christ is resurrected from the dead, some of the disciples have to see it; the scriptures do not suffice. They had not understood the word, which has to be visualized and incarnated as the Word itself.[6] Saint Thomas was not the only one to practice this rational skepticism, in a certain sense a premodern epistemology. To see is to believe, but above all to confirm, to know, or to recognize.

None of this contradicts what was said in the previous chapter about the weight of the word in the ancient world, since the scholastic reading of Aristotle was dominated by rhetoric and logic, and the Augustinian reading of the Bible by the bliss of those who believe without seeing, the truly blessed, the elect. In medieval theology, knowledge had to serve faith, as the visible had to serve the invisible, the earth the heavens, and the present the past. In the world of knowledge before the age of modern science, things and their visual representation had to match words.

The modern world altered these relations. At first an attempt was made to adapt natural facts to the written authorities (the rhinoceros to Pliny and Strabo), but soon the opposite strategy was applied: words had to be adapted to the facts of that other great book, the Book of Nature, whose language seems to be mathematical. The modern world altered the relationship between word and image, displaced the invisible, and granted unlimited value and credibility to what the eyes could see, as long as they were duly trained to observe and assisted by the appropriate instruments. In fact, images not produced by the naked eye would end up exercising hegemony in the world of science.[7]

It is significant that at the beginning of the seventeenth century one of the first European scientific academies should call itself the Accademia dei Lincei, a tribute to the highly acute vision of the lynx, whose name was also associated with a mythological character who could see ships from a great distance. If science, as William Eamon has argued, making use of this motif, was understood in the early modern era as a hunt, an activity associated with the ambush and capture of prey,[8] it seems obvious which human faculty and sense would be the most important: sight, the precise observation of forms, colors, light, and movement.[9]

As modern science took shape, the pictorial representation of natural phenomena came to assume an unprecedented relevance. The introduction of perspective to Western art, a development in which Dürer was a principal actor, is the key to understanding not how one generation after another learned to represent space and objects, but how they learned to perceive them.[10] In much of his work, Dürer similarly demonstrated his mastery of the new emphasis on achieving a fit between the seen and the represented, a correspondence between the original and the portrait, what we call *verism* in painting and could more generally call *realism*.

It is realism that distinguishes his drawings, prints, and paintings of fauna: that hare whose whiskers and gaze have an almost photographic presence; the elegant anatomical study of the wing of a blue-bellied roller; the texture of the surface of a crab; or the expression of a walrus. And what are we to say of one of the most famous Western paintings of flora: Dürer's watercolor of a group of plants, *The Large Piece of Turf,* in the Albertina museum? More than a magnificent botanical illustration or a masterpiece of realism, there is something hyperreal about it *avant la lettre.* It is not only the minute attention to detail, the accuracy with which Dürer represents variables of common and identifiable species: the grasses, the yarrow, or the fiorin (*Agrostis stolonifera,* to give it its scientific name). Neither is it just that he captures the air, the movement, the stalks of the dandelions bent by their own weight. It is that he manages to reflect the organizational disorder that is nourished from the subsoil, the spontaneous groups formed by the different shapes and species—in short, the chaos of life (Figure 8).

Art and nature confront one another as rivals—just like an elephant and a rhinoceros—to see which is the stronger. There is no need to think of Dürer's self-portrait as Christ; these representations of nature alone are sufficient to support the claim that Dürer is giving the Creator a shove, playing at creating a work that imitates and surpasses the divine creation. His objective was to reproduce it. For Dürer, this meant following the

Figure 8. Albrecht Dürer, *The Large Piece of Turf*, 1503 (Albertina, Vienna).

practices of the master jewelers and silversmiths whose long apprenticeships required them to imitate both their master and nature itself. Apprentices spent hours and hours copying objects and real phenomena from life. The observation and imitation of nature were also crucial for the painter, as he recalls in the aesthetic excursus of his treatise on proportions:

> But life in nature manifests the truth of these things. Therefore observe it diligently, go by it and do not depart from nature arbitrarily, imagining to find the better by thyself, for thou wouldst be misled. For, verily, "art" [that is, knowledge] is embedded in nature; he who can extract it has it.[11]

Dürer was also a major precursor in the field of landscape painting. The watercolors that reflect his travels through the Rhineland and the Low Countries, the Alps and the Tirol are among the first works of Western art to portray ordinary locations that are perfectly recognizable. Dürer thereby conferred artistic dignity not only on those simple villages and geographical formations, but also on the technique itself, the watercolor, which until then had been regarded as a minor art. Both his renderings of fauna and flora and his landscapes are innovative and forward-looking for two reasons: for their subject matter and for their technique.[12] What is at stake in each case is the realism achieved by attention to detail, by training the eye to observe the proportions of objects, by the knowledge of optical laws, by the ability to apply geometry to the handling of volume, and by meticulous observation of the effects of light on bodies and space.

Dürer attests to his concern for verism in his captions, his field notes, the comments in his diaries, and his theoretical writings: "And thou must know, the more accurately one approaches nature by way of imitation, the better and more artistic thy work becomes."[13] It was the weight of testimony, what is known as *ad vivum* ("from life") in painting, and found its correlate in the empiricism of the natural sciences, which would lead to experimental philosophy.[14] Both practices are consolidated in the Western tradition in the same period. Realist epistemology, which asserts the supremacy of the direct experience of the natural world by the senses as the most reliable and legitimate basis for its representation, is a phenomenon that in retrospect can be seen to have developed simultaneously in science and art. This parallel consolidation of figurative art and the new science is no coincidence.[15] Neither is the important, though at times underestimated, role in both fields of the craftsman, whose skill was based on direct observation, imitation, technical experimentation, precision, and contact.[16]

One of the most famous paintings in Britain's National Gallery, the *Arnolfini Wedding*, contains the words "Johannes de Eyck fuit hic 1434." It says the artist, de Eyck, was here: there could be no better way to assert the testimonial power of an eyewitness. The same is true of Dürer's natural history engravings. The monogram AD, repeated in his works, seems to say "Albrecht Dürer was here, saw this, and copied it." The presence of the artist, the direct observation of the phenomena represented, confers legitimacy on the observation, guarantees its verisimilitude, and raises the status of both the artist and the work of art. The mastery of the artist is determined by the exact correspondence between phenomenon and representation, while the reputation of the artist guarantees that the representation is a faithful and exact imitation of what he has seen with his own eyes.

ALL THE SAME, Dürer never saw the rhinoceros that he immortalized. We know that he had access, in the house of the scholar Conrad Peutinger, to a letter from Valentim Fernandes, and no doubt to the accompanying sketch: the two testimonies that hold the key to the mystery of this curious history. Dürer's rhinoceros is thus the product of secondhand testimony. It is the result of an absence, not a presence. Nothing would have been possible (visible) without the aid of a powerful imagination, since to represent what you have not seen yourself requires not only direct experience of other relevant objects or representations, but above all artifice and fiction.[17]

Conrad Peutinger (1465–1547), a friend of Dürer and of the Emperor Maximilian, was a great humanist, an extraordinary bibliophile, and an expert in classical languages and antiquities. In fact, one of the possible sources of inspiration for Dürer's engraving that is often mentioned is the existence of several Roman coins with the image of a rhinoceros, which Peutinger may have had in his numismatic collection. As for Valentim Fernandes, this Moravian printer and typographer arrived in Lisbon in 1495 and was to spend more than twenty years there. As a translator he produced Portuguese versions of such works as the *Travels of Marco Polo* and Andrés de Li's almanac *Repertorio de los Tiempos*, but Fernandes was above all an expert on maritime expansion. He produced a relatively well known *Islario* and kept certain humanists, artists, and German notables well informed about Portuguese discoveries. He corresponded regularly with Peutinger, whose private archive contains numerous unpublished manuscripts and reports by Fernandes on matters of this kind.

Nuremberg, the city of Dürer and Peutinger, was bustling with news at that time as it was an active crossroads for trade, science, and the arts, and the center of the German Renaissance. The ties between Nuremberg and Lisbon were close, perhaps surprisingly so for a modern reader, although, paradoxical as it may seem, Europe existed long before the nations that form it today. The relations between Portugal and the German Empire involved exploration, astronomy, and financial capital. German merchants had set their sights on, and committed their money to, the new routes and businesses that the Portuguese navigators were opening up. The Fugger family, for example, had an agent in Lisbon, where there was a sizable German colony at the end of the fifteenth century.[18]

In the 1470s the astronomer and mathematician Johannes Müller von Königsberg (better known as Regiomontanus) established his observatory and his own print shop in the city of Nuremberg. Thanks to printing, his commentaries on the Ptolemaic system and his contributions to the field of trigonometry (including many tables and almanacs, images and diagrams, drawings of instruments and planetary movements) traveled widely and rapidly and reached a good number of customers in far-off places within a decade or two. Among them were Amerigo Vespucci and Christopher Columbus.

Martin Behaim, a native of Nuremberg, was a pupil of Regiomontanus. His life reveals the ties between Portugal and Germany and how news of the geographical discoveries was making rapid headway in the workshops of the city. Besides being a geographer and mathematician, Behaim was a merchant. He visited Lisbon for trade, where he ended up as a member of the Junta dos Mathematicos, a council that advised King John II on navigation. After Behaim's return to Nuremberg some years later, the authorities in his home town commissioned him to make an object unique in the history of globalization. This was the terrestrial globe, the first of its kind, which was made between 1490 and 1492. It was profusely annotated with captions identifying the products, the routes, and the middlemen in the spice trade. Ferdinand Magellan used it to gain financial support for his venture and, like Columbus, first offered his project to Portugal before turning to Castile. Magellan's circumnavigation (1519–1522) was largely financed by German capital.[19]

Without forgetting Venice, the traders of Nuremberg were putting their bets on Lisbon, the port through which the East was beginning to disembark in Europe. The terrestrial globe, a masterpiece of the age of discovery, is an artifact that heralded and secured the unification of the world, but it is also an object that illustrates how technology, commerce, and science were intertwined in Nuremberg. The word *Kunst*, which we have seen related to knowledge *(scientia)*, also means an artifice constructed by a craftsman—a human product like the products of nature such as plants, quadrupeds, or stars. In our contemporary vocabulary, *kunst* means art: some words, like some artifacts, and not just German ones, have the capacity to sum up and concentrate everything.[20]

Dürer's rhinoceros is also a memorable chapter in the relations between Lisbon and Nuremberg, as well as a story about the circumnavigation of the globe and the artistic and technological means by which that feat was accomplished. For an artist interested as he was in representing the natural world, fauna from distant lands had an added attraction because, like other humanists, the German master had a strong fascination with the exotic.[21] It is the thirst for novelty and the remote—shared by Isabella d'Este—that makes the vocation and task of the artist like those of any other researcher. It is symptomatic that in so many languages the semantic field of curiosity is associated with desire and novelty, inquiry and experiment (*Neugier, Suche,* and *Versuch* in German, for example). Dürer, for instance, was very interested in the jewels and feather work of the Aztecs, which he was able to admire in Brussels when Charles V exhibited the American treasures sent by Cortés in the summer of 1520. He noted in his diary, "In all my life I have seen nothing which has gladdened my heart so much as these things. For I have seen therein wonders of art and have marveled at the subtle ingenuity of people in far-off lands."[22] It was on that same journey, undertaken to attend the coronation of the emperor, that Dürer contracted the malaria that would plague him for the rest of his life after visiting a mosquito-ridden part of the coast of the Low Countries in the vain hope of seeing a beached whale—a fate resembling that of Pliny the Elder, whose eagerness to witness the eruption of Vesuvius as closely as possible cost him his life. Direct contact is not without its hazards; to seek experience is to expose oneself to risks.

We can imagine how much Dürer was fascinated by what he read and even more by what he saw in those documents in Peutinger's library in the summer of 1515. The shape and nature of the rhinoceros must have captivated him at once. Just as poets want to give something a name for the first time, to confer names on the world, painters dream of representing something that has never been seen before. It was an important moment, then, when the rhinoceros sent from the Indies to fulfill the political ambitions of Manuel I, which aroused the curiosity of erudite humanists and the mundane spectators of Lisbon and died en route to serving the courtly tastes of Pope Leo X, arrived in Nuremburg in the form of a simple drawing set before the eyes of one of the most powerful image-makers ever.

Dürer lost little time in reproducing it in the drawing that is kept today in the British Museum. It is a preparatory work for the woodblock to be cut by Hieronymus Andreae, the engraver in Dürer's workshop at the time, or perhaps even by the artist himself, since he was a consummate expert and enjoyed wielding a gouge and knife with his own hands to make incisions in the block.

It is obvious that Dürer's sketch of this extraordinary animal is copied from or inspired by an image. Valentim Fernandes's original letter has been lost, but there is a copy (in translation) in the Biblioteca Nazionale in Florence.[23] What has been lost without trace is the drawing that must have accompanied the letter. In any event, the image created by the master could never have been based on the textual description of the rhinoceros—the *ekphrasis*—alone. The letter repeats the accounts of Pliny and Strabo, describes the arrival of Ganda and the duel with the elephant in Lisbon in very similar terms as the version by Damião de Gois, and then continues with a few remarks on the geography of India and its different regions, with details of the Portuguese movements and conquests in that part of the globe. The morphological description of the animal focuses on what had by now become the usual clichés: its resemblance to an elephant in size and to a bull in shape; the presence of a horn on its nose; its two plates; coils or folds like those of a dragon, serpent, or other reptile (*duo cingula tamquam draconum volumina*). This is clearly insufficient to produce the famous image.

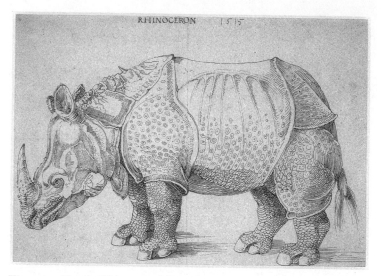

Figure 9. Dürer, *Rhinoceron 1515*. Pen and sepia ink drawing.

A lost sketch by Valentim Fernandes, or more likely of a Portuguese collaborator of his, guided that of Dürer. It has been suggested that the pen and brown ink drawing in the British Museum is not by Dürer, but is itself the image that accompanied the Moravian's letter.[24] This is unlikely: the watermark of the leaf confirms the authorship of Dürer's workshop (Figure 9). Moreover, the text that appears below the drawing—the text that will be copied with slight modifications and placed in the upper part of the engraving, just one of the elements in this history of originals and their copies—was written by Dürer, though this is not his handwriting; it is a transcription, copied from a lost original, and reads:

> On 1 May 15[1]3 this animal was brought to our King of Portugal in Lisbon from India. It is called a Rhynocerate.

I send you this drawing because the animal is very surprising. It has the colour of a toad and is covered all over with thick scales, and in size it is as large as an elephant, but lower, and is the deadly enemy of the elephant. It has on the front of the nose a strong sharp horn: and when this animal comes near the elephant to fight it always first whets its horn on the stones and runs at the elephant with his head between its forelegs. Then it rips the elephant where its skin is thinnest and then gores it. The elephant is greatly afraid of the Rhynocerate; for he gores every elephant he meets. He is well armed, very proud and alert. The animal is called Rhinocero in Greek and Latin and Gomda in the Indian language.[25]

Apart from the mistaken date, which would be repeated in the caption to the engraving, Dürer transcribes literally the commentary that was appended to or included with the sketch done in Lisbon, the only portrait done *ad vivum*, or at least after its mysterious author had seen Ganda with his own eyes. As Fontoura da Costa pointed out in 1937, the author of the reference to "our King of Portugal" cannot have been Dürer himself.[26] In any case, the words "I must send this representation" dispel any possible doubt: Dürer was transcribing words that were not his own. He copied what he read.

And what did he see? This has been the crux of the matter for many. We will never know. The question of to what extent Dürer's rhinoceros (the drawing rather than the engraving) is or should be attributed to someone else is legitimate but misguided. It is like asking whether Columbus was the first European to reach America or not, highlighting the question of priority and singularity, while playing down the social and cultural impact of the discovery. Columbus's voyage unleashed a process infinitely more decisive and complex than the mere fact of having disembarked on a distant shore. The incorporation of the New World into the geography of the Old, the expansion of the notion of the inhabited universe, the forging of a shared history and Atlantic networks, and the long series of questions, debates, and problems of an anthropological, political, and scientific nature that have arisen from that fateful voyage in 1492 for the people on both sides of the Atlantic cannot be compared with what a few improbable Irish monks or seafaring Vikings achieved or brought back many years earlier. Without communication, without a public dimension, a discovery or an invention is nothing. Science is a profoundly social practice.[27]

SO IS ART. Albrecht Dürer was one of the first to fully understand its significance. The woodcut *Rhinocerus 1515* illustrates this abundantly. When the artist designed it, he was at the peak of his career and his fame as a master engraver. He had explored the potential of the woodblock further than anyone else and had taken experimentation with engraving, etching, and drypoint along untrodden paths, creating genuine masterpieces in both wood and metal.

He was the third of eighteen children of a well-known Hungarian master goldsmith. His godfather, another worker in precious metals who lived in the same street, was a baker's son who abandoned working in silver to dedicate himself to new activities as a printer, engraver, and publisher in the same year in which Dürer was born, 1471. He was Anton Koberger, the founder of the first print shop in Nuremberg in 1470, where three years later he published the *Nuremberg Chronicle*, one of the first incunabula and a gem in the history of the printed book. Published simultaneously in Latin and German, the *Chronicle* contained 1,809 woodcuts made in the workshop of Michel Wolgemut, another craftsman entrepreneur who experimented and opened businesses in several of the visual arts. These illustrations boosted sales of the book and had a life of their own as printed sheets.

Dürer began his training in his father's workshop at the age of twelve. In the following year he demonstrated his enormous talent with his first self-portrait, an introspective work that affirms the artist as he explores and copies his own image. Besides being one of the most precocious and delicate self-portraits in the history of art, it is worth noting here that it was done using silverpoint, a technique that allows no margin for change or error. His skill at such an early age in a technique like those used in the family metalworking business, a trade that was developing rapidly in Nuremberg at the time, explains why his father realized that the boy was not going to confine himself to making pyxes, reliquaries, and caskets for the rest of his life. In 1486 he sent his son to study in Wolgemut's workshop for three years, where he probably helped the draftsmen and engravers in preparing the blocks to print the illustrations of the incunabulum.[28]

Then came the young Dürer's journey to Basel and Strasburg, where he assisted with the illustrations for various editions. After returning to Nuremberg, he married and established his own print shop in 1494. His business was modeled on that of Wolgemut, then a pioneer in establishing alliances with businessmen like Koberger, who in turn managed to become the most important printer in the whole of Germany. Dürer would have learned from both that the ideal was to control the whole process as much as possible and to manage personally the different aspects of this burgeoning industry, from design and engraving to printing, publishing, distribution, and sales.

Dürer rapidly understood the potential importance of printing and serial production both personally and more broadly in art and commerce. He found the autonomy that he could achieve through new forms of work attractive. If he managed to sell his engravings on a large enough scale, he might be able to dispense with patronage, the usual form of subsistence and livelihood for any artist of the era. And to a large extent that is what happened: his woodcuts and engravings gave him liquidity and independence. He could not dispense with patronage altogether, though, and continued to seek the support of nobles (including the Emperor Maximilian), but he did not depend entirely on them—fortunately, as not even Maximilian was a punctual payer.

To control his engravings from the design to the production, distribution, and retail stages meant setting up an infrastructure and hiring specialists. The artist was turning into an entrepreneur. Dürer managed to find agents in distant cities to put his work on the market. It can be stated without any risk of exaggeration that he contributed decisively to the creation of a new profile of the artist, one that combined the tradition of the medieval craftsman with the new worlds of the erudite humanist, the man of science with an interest in geometry and optics, and the bold entrepreneur characteristic of the emergent bourgeois culture.

When the rhinoceros came into his hands, Dürer had already achieved wonders in the field of engraving.[29] One of his first series of woodcuts, illustrations for the *Apocalypse* (1496–1498), had already become a point of reference for religious iconography and was soon to become the most influential set of images in central Europe during the period of the Reformation. He published it again years later, when he created two other extremely successful series of woodcuts: scenes from the *Life of the Virgin* and the *Large Passion* (1503). Those three series (*Apocalypse, Life of the Virgin,* and *Large Passion*), known as his three major books, were illuminated with woodcuts, by then a well-established technique in Nuremberg. Less expensive than metal engraving, it was more suitable for book illustration for the simple reason that the wood blocks had the same thickness as the movable type of the press, which made it easy to combine images with text.

As for the use of the burin, his cutting tool, Dürer had experimented and brought his technique to an astonishing degree of perfection. It was a technique developed in the workshops of the silversmiths, where engraving was more difficult, but held out more possibilities than did woodcarving, as time would make clear. Although metal engraving developed later than woodcarving, the improvements in the technique that were introduced in the course of the sixteenth century led to its progressive substitution for the woodcut in Germany. By the seventeenth and eighteenth centuries the woodcut was confined to popular, inexpensive works, at least in book publishing. Dürer had engraved metal sheets with a burin to produce images like *Nemesis (Great Fortune)* in 1502 and *Adam and Eve* in 1504, long before reaching the zenith of his creativity with the trilogy of masterpieces of engraving with the burin in the years immediately preceding the rhinoceros: *Knight, Death and Devil* (1513), *Saint Jerome in His Cell* (1514), and *Melencolia I* (1514), one of the most famous engravings of all time, or at least of our own, thanks to the classic study by the historian of philosophy Raymond Klibansky and two historians of art associated with the school of Aby Warburg: Erwin Panofsky and Fritz Saxl.[30]

With all this experience and this oeuvre behind him, Dürer tackled the execution of the figure of the rhinoceros with something for which there is only one word: mastery. It is a resolute image, as powerful as the pachyderm itself. It reveals what has been attributed so many times to its author: *Gewalt*, a creative vitality and determination, drive, authority. In spite of the splendid detail of those plates, handled as though they were metal armor, and the scaly, cracked skin of a millenarian reptile on its legs that stick out from the carapace, the artist does not appear to have lingered over the work, but moved with a brisk confidence. The severe lines and voluminous shape of the huge pachyderm seem to correspond to a firm, secure, more or less rapid execution.

The two most striking features of the animal's prodigious morphology—the horn and the hide—dominate the image to the extent of almost overshadowing the rest. As we have seen, Ganda was little more than that to spectators: horn and armor, to gore and to resist. You have to look at the engraving for a long time before you notice that half-human eye with its melancholy, tired expression. As for the horn, it is a rocky protuberance that emerges from an earthy, vegetable-like snout. If it were not for the eye that animates the beast, there is something about his front that makes it appear materially indeterminate, situated somewhere between the organic and the inorganic, amid the lichen, moss, and stones. It is evidently a prehistoric animal, and a legendary one: the second horn that juts from its back may have been a slight deformation of the skin, some kind of cyst or protuberance, but it has usually been interpreted as a

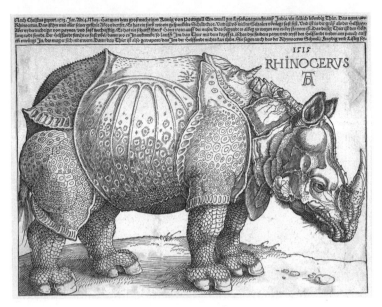

Figure 10. Dürer, *Rhinocerus 1515*, woodcut.

remnant of the mythical unicorn, the oriental *monoceros* that precedes and prefigures it. It is an image in which evidence of the Asian pachyderm and the imagined anatomical features of the unicorn and of reptiles, the other natural enemies of the elephant, cohere seamlessly (Figure 10).

The chain-metal armor evokes both a dragon and a reptile and, at the same time, the knight or samurai who has to combat it. This is where Dürer situates and transmits all the force of the animal and where he simultaneously displays his great debt to his experience designing armor (an important facet of his artistic production), his familiarity with drawing, and the flexibility of the overlapping surfaces.[31] Without a shadow of doubt, it is a chimerical rhinoceros, made up of segments of tortoise, unicorn, and dragon, halfway between fact and fiction, an image from a sample book of imaginary zoology from which flows a current of epithets, legends, and words.

It is not by chance that the caption is set above the image and dominates it. The image appears below the weight of the written word, to which it is subordinated. This hierarchical distribution of word and image establishes a curious balance between what can be read and what can be seen, the old dialectic that permeates Western thought and can be understood, as Leonardo understood it, as a *paragone* between the two forms of representation.[32] There is a struggle, a certain tension, a rivalry between them both. Which will emerge victorious? Which best reproduces and conveys phenomena, objects, the world?

Both the words and the image have been modified in the engraving. It is not the same image as the one in the drawing, and the caption has been modified too. It now runs:

> After Christ's Birth, in the year 1513, on 1 May, this animal was brought alive to the great and mighty king Emmanuel at Lisbon in Portugal from India. They call it Rhinoceros. It is here shown in full stature. Its colour is that of a freckled tortoise, and it is covered by a thick shell. It is the same size as an elephant but has shorter legs and is almost invulnerable. On the tip of its nose is a sharp, strong horn which it sharpens itself against stones. This animal is the deadly enemy of the elephant. The elephant is afraid of it because upon meeting it charges with its head down between the elephant's front legs, and gores its stomach, before the elephant can defend itself. It is also so well armoured that the elephant cannot harm it. They say that the Rhinoceros is fast, cunning, and daring.[33]

Perhaps the most significant difference (besides the obvious error of the date: 1513 instead of 1515), is that the text that accompanies the engraving, the image that is going to circulate and make the oriental prodigy known all over the world, omits its own name, Gomda or Ganda. This omission lends itself to a postcolonial reading. To give things names, to represent them, to rename them, or to alter the toponomy or names of rivers or living beings—these are not innocent acts. Instead, they order and determine the world.

The text is essential for an understanding of the image, whose irresistible power lies in the way it reinforces and makes visible what is depicted, what is intangible but present, what is imaginary but very real. Dürer's rhinoceros continues to fascinate because it opens a window on a world that has disappeared, a world animated by the remarks of Strabo, the stories of Pliny, and the humanists' expectations of an imaginary Orient. We too are brought closer to something that is remote, lost, and therefore desirable. It is, moreover, yet another demonstration of the poetic function of language, since words provoke the imagination, producing images and making things visible, just as pictures claim to do. And like images, of course, words are read and tell stories.[34] There is no clear-cut boundary between the legible world and the visible world, just a dense network of arteries that communicate them and ensure a flow of traffic.

In a certain sense Ganda disappears into thin air at the moment Dürer's rhinoceros stimulates the imagination. In the same year, 1515, in neighboring Augsburg, one of Dürer's assistants and a master engraver like himself produced another woodcut of the same animal. Only one copy is known (in the Albertina, Vienna). It was made by Hans Burgkmair (1473–1531) during the same months and has given rise to all manner of

speculation about its relation to Dürer's print, since the Augsburg master, creator of more than half of the illustrations for
The Triumph of Maximilian and an innovator in the use of
chiaroscuro in woodcut, was even closer to news about the Portuguese discoveries than Dürer was. Burgkmair belonged to the
small circle of artists and merchants in southern Germany who
had interests in the Orient via Lisbon. In the previous decade, for
example, he had provided images of Oriental fauna to illustrate
the report on a voyage by German traders to Africa, Arabia,
and the East Indies by Balthasar Springer. These woodcuts
may have been based on drawings done *ad vivum* in India.[35]

It is also probable that this rhinoceros was based on the same
sketch that inspired Dürer, the lost original of Lisbon. In any
case, it is fortunate that this second engraving has been preserved as it is more realistic, more faithful to what an eyewitness would have seen, than the other (Figure 11). It lacks the
textual commentary and armored appearance of Dürer's animal; it has no legendary metallic defense. It fails to pass on the
fables and words of antiquity. We observe this rhinoceros and
appreciate its domestication. This is Ganda: far from seeming
invulnerable, it is presented to us as a subjugated beast. Its front
feet are bound by a thick rope that reminds us of its real condition as a captive. Ganda is no fearful and fantastic beast, but a
tortured and humbled colossus.

REALITY, HOWEVER, DOES not always match our desires
and our dreams. Burgkmair did not reach the artistic heights
of Dürer in his field. As creators of images, they cannot be
compared. Burgkmair could not compete with Dürer in a
more prosaic and mundane sense either: the Nuremburg master

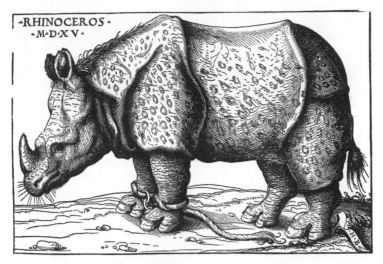

Figure 11. Hans Burgkmair, *Rhinoceros 1515*, woodcut (Graphische Sammlung, Albertina, Vienna).

enjoyed much more success in the market in his lifetime, and the demand for his engravings has endured. Albrecht Dürer was copied, collected, reproduced, and resurrected from the dead to such an extent that art historians talk about a genuine revival of his work in central Europe between 1570 and 1630, only a century after his birth, particularly in Antwerp and the courts of Bavaria and Prague, where Rudolf II managed to bring together more than four hundred original Dürer drawings.

Works of art sometimes undergo the same fate as scientific discoveries. Through their diffusion all over the world and across the generations, they become points of reference in their field. We take their existence for granted and lose sight of the very specific material, technological, and social mechanisms that were crucial to their creation and also to their fame, to the colonization of gazes and minds through which certain cultural products become authentic classics.

While the model of serial production guaranteed Dürer the freedom to create and manage his work, as well as the exercise of a certain monopoly over the latter, his collection of woodcuts and engravings and their extraordinary dissemination in space and time granted him and his oeuvre an unprecedented visibility and universality. This explains the triumph of Dürer's rhinoceros over that of Burgkmair. In fact, it explains its extraordinary hegemony over all other images and representations of the legendary animal. What is more, it explains its power over reality itself, when European travelers arrived in the Indies or when other live rhinoceroses, the successors of Ganda, were brought to Europe. The centuries-long sovereignty of the words of Pliny was replaced by the influence of an image generated precisely in order to be repeated and reproduced endlessly. The art of engraving was a technology made to rule the world and to domesticate the gaze.

The crux of the whole issue is contained in a single word in the caption: "Das ist hie mit aller seiner gestalt *Abcondertfet*." J. M. Massing, whom we follow here, translates this into English as "It is here shown in its full stature." Others translate it as "This is an accurate representation" (see Wikipedia's entry on Durer's rhinoceros, for example). A stricter translation of "Abcondertfet," attentive to etymology if not to modern meaning, might be "It is here *counterfeited*." *Abcondertfet* is derived from the Latin *contrafacta*, "taken from, or in opposition to, something else," which expresses all the complexity and ambiguity of the act of representing something *ad vivum*. As in English, in which a counterfeit is a fraudulent imitation, but one copied accurately from life, the German *abconterfeiten* and its cognates display the same polysemy. It is connected with a decidedly old word from the thirteenth century that has the sense of falsity *(gunderfray)*, although in the sixteenth century its use is extended to include the act of portraying *(porträtieren)*, imitating *(nachmachen)* or reproducing *(abbilden)*. In fact, the form adopted by Dürer, which is closer to Latin than to German, was soon to be used as a substantive to designate a portrait done from life *(ad vivum)*, which has led to all kinds of readings and interpretations.[36]

And what is an image if not precisely that, a copy and a portrait, simultaneously a reflection and a falsification of reality? The Latin *imago* and the Greek *eikos* refer to resemblance and similarity, a notion with more of a spiritual than a formal meaning in both Platonic and Christian tradition. For the Romans an *imago* was also the portrait that was carried in aristocratic funeral processions as well as a generic term to refer to the ancestors.[37] An image is a phantasm, a double, an illusion of reality. When God created Adam in his image and after his likeness, or when the prisoners distinguished the shadows in Plato's cave from reality, we are dealing with a resemblance to the model that is not sensory, physical, or material. The imitation of appearance, anatomy, or morphology is no more than a reproduction or a falsification of things. In this sense, *Rhinocerus 1515* is an image that not only imitates what can be perceived by the senses, but is also charged with preconceptions and mental resonances. Perhaps its power resides in how it materializes them.

We know that it is not a portrait or a reflection, and yet at the same time that is what it wants to be; that is what it simulates in its appearance. It presents itself as something authorized and legitimized. It is called a portrait of its form *(Gestalt)*, but it is much more: it is a work of art in the prehistory of the age of mechanical reproduction, as Walter Benjamin formulated it with film and photography in mind rather than engravings.[38] In a truly modern manner, the copy takes the place of the original. The aura of the unique and the singular gives way to fascination with the copy and the artist. Thanks to art and technology, what was originally an exclusive diplomatic gift, a living creature worthy of monarchs and even of the pope, becomes a product that reaches the masses and is produced for the masses. In the sixteenth century even a baker, a tailor, or an ironsmith could own it. It is the triumph of reproducibility and exhibition, whose apotheosis is celebrated in today's museum shops where the item is made accessible to all.

Dürer's rhinoceros (which will soon be infinitely more Dürer than rhinoceros) will encircle the globe made by Behaim, itself a reproduction (imitation or falsification) of the original but one that enabled Magellan and Juan Sebastián Elcano to circumnavigate the world. The armored unicorn will be able to do so thanks to an invention as revolutionary as the wheel, an artifice in which is played out the possibility of replication, deceit, and knowledge.

It is more than fifty years since William M. Ivins, then curator of prints and engravings at the Metropolitan Museum of Art in New York, published a marvelous study on prints and visual communication.[39] Its importance was picked up by Marshall McLuhan, the prophet of globalization, whose argument has been revived more recently by Bruno Latour. Ivins argued that the cultural impact of print on the scientific world was driven less by the written word than by the dominance of the image and the new capacity to reproduce drawings and silhouettes, making it possible to identify or imitate things—recognize a plant, erect identical fortifications, make two identical instruments—at distant points on the planet.

Ivins's scholarly work began with the testimony of Pliny, whose *Natural History* narrated the diffuse origins of the pictographic representation of natural phenomena, before going back to the problems encountered by the ancient Greeks when faced with the root problem behind all these concerns: How is one to universalize the name or the form of a plant? How is a consensus to be achieved on a species that will enable the members of a community to guarantee that they are referring to the same thing? Without such an agreement, any scientific progress can be ruled out. Nomenclature cannot save the day, because all creatures and species have different names in different languages and places. Needless to say, verbal language is the most conventional and social language. The path to be followed had to be that of visual representation, the production of an identical image, a faithful and exact copy of the thing itself. Difficulties already arose when it came to producing faithful images of things, because that depended on uniformity of perception (visual production too is plagued by conventions, and pictorial language is just as much an ensemble of arbitrary, socially legitimized signs). The problem also consisted in something quite simple, but which precisely for this reason became an insidious obstacle: the reproduction of two identical images was out of the question in the ancient world.

With the development of the technique of the woodblock in the middle of the fifteenth century, an invention overshadowed by Gutenberg's movable type, the West could finally escape the shackles that had restricted all Greek philosophy, Roman engineering, and scholastic thought. The first books to contain woodcuts that were not just edifying or decorative, but clearly informative—in other words, openly aimed at producing and reproducing knowledge—were produced between 1470 and 1500: treatises on war like *De re militari* of Valturius (1472); books on astronomy such as the almanacs of Regiomontanus (1474) or the *Sphera Mundi* of Johannes de Sacrobosco (1485); travel books like the *Peregrinationes* of Breydenbach; and of course some central European herbals such as the *Gart des Gesundheit* (1485) and similar works that, decades later, were to lead to classics of botanical illustration such as Otto Brunfels's work *Herbarum vivae eicones* (1530) with woodcuts by Hans Weiditz, and the *De historia stirpium commentarii insignes* (1542) by Leonhart Fuchs.

It is in the latter that a noteworthy phenomenon occurred that facilitated communication but tended to eliminate the importance of direct testimony. It was something that favored the transmission of knowledge at the expense of the experience of the senses. Those illustrations emerged from an alliance between the work of a physician or apothecary and the work of a draftsman who accompanied him on his travels or who was advised by him. They began by reflecting images of particular, individual species drawn from life. However, a problem soon arose because what is specific is sometimes (very often) an obstacle to recognition of the universal. Fuchs, for example, chose to represent a diagrammatic and recognizable form of the species rather than an individual example. When all is said and done, what was a scientific image supposed to represent? A specific case or a case of universal value?[40]

The herbal that Fuchs published in Basel included species described by ancient authors, local species, and yet others from the Americas. They were arranged alphabetically and were accompanied by illustrations. The images were schematic and avoided shadow or volume (unlike those of Brunfels). They presented the generic form of the plant in silhouettes and lines. The objective was to aid the identification and observation of the plant. Originally published in Latin, the work was issued in the vernacular in the following year, and soon afterwards in small octavo volumes, reducing the text to its minimal expression, with the names of the plants in several languages and the images cut down to a smaller size. The obvious convenience of such a smaller edition would ensure it a much wider readership, which is what happened: it became a bestseller in France, Switzerland, Germany, and Flanders.

According to Ivins, a lot could be learned about the history of knowledge from those illustrations, some of which appear infantile or even amusing from a present-day perspective.[41] He was right. The parallels with Dürer's woodcut are striking: just as the readers and botanists who came after Brunfels and Fuchs carried out their fieldwork by trying to recognize the species illustrated in those books (a procedure that would be practiced more systematically in the second half of the eighteenth century by the followers of Linnaeus, when globalization was hastening to a close), the idealized version of Dürer's armored rhinoceros ended up as the model for recognition and imitation, all the more so because of the rarity of the animal and the dearth of possibilities for comparison with the original or with other images. All this, plus the fact that Dürer never actually saw the rhinoceros, has attracted the attention and elicited the remarks of many historians, including some great ones such as Ernst Gombrich, who included it in his *Art and Illusion* as a paradigmatic case concerning truth and stereotype.[42]

No fewer than 45,000 copies of the 1515 engraving were sold during Dürer's lifetime. There were two further editions in the

1540s, which had even more impact on the market. Before the century was out there were another two editions in Germany and two in the Low Countries, all deriving from the original block, which was progressively deteriorating. W. Janssen's chiaroscuro edition was published in various colors in 1620, the most widespread of which is probably the one in olive green popularized by the British Museum.

While it is true that Dürer's rhinoceros soon became more Dürer than rhinoceros, the opposite is also true, since it rapidly abandoned the master of Nuremberg's woodblock and ceased to belong to him. It was Dürer himself who initiated this reproductive task by incorporating it on a smaller scale in a detail of the *Triumphal Arch of Maximilian* (1517), but it soon slipped out of his hands. It already features in a *Book of Prayers* of Maximilian (1520), a silhouette marked by Dürer's image, a rhinoceros more heavily armored than the one in Manuel I's 1517 *Book of Hours*. Along with the successive editions of the engraving, within a few decades there were outstanding books that repeated that figurative anatomy. Sebastian Münster placed it on one of the pages of his *Cosmographiae* (1544), and Conrad Gessner included it in the first edition of the *Historia animalium* (1551–1558). Both were tremendously influential scientific works. Münster's went through more than twenty editions before the end of the century and presents an interesting detail in the history of our image: the engraver, David Kandel, an artist known for his natural history representations, wanted to leave his initials beneath the belly of the rhinoceros to replace those of Albrecht Dürer above it (Figure 12). This is the history of a copy, or rather of a repeated succession of copies, but it is also the history of artists in search of fame and prestige, profits, and

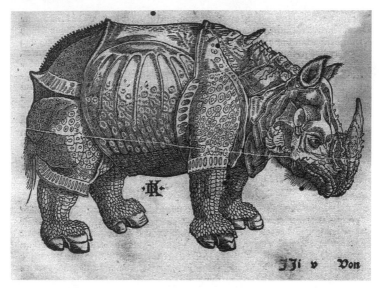

Figure 12. David Kandel, *Rhinoceros*, engraving in Sebastian Münster, *Cosmographiae*, 1544.

a taste of immortality, starting with Dürer and continuing with the rest. The monograms succeeded one another as the artists repeated one another like the shapes and lines of our armored pachyderm.

Gessner's zoological encyclopedia in three volumes was also reissued in the second half of the century and in the course of the following one without any break in continuity, which enabled Dürer's rhinoceros to permeate the whole of early modern natural history (Figure 13). The image that appeared in this work was a deliberate reproduction of Dürer's engraving and recognized his authorship in the text. The image passed from Münster and Gessner to the *Cosmographie Universelle* of André Thevet (1575) and to the *Discourse de la mumie, de la licorne,*

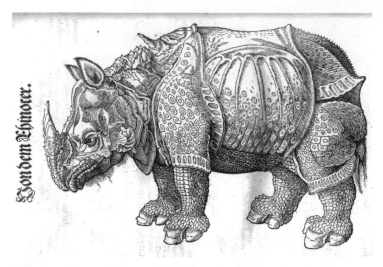

Figure 13. Conrad Gessner, *Historia animalium*, 1551–1558.

des venins et de la peste (1582) by Ambroise Paré, physician and surgeon to the French court and an essential author for the anatomy and teratology of the period.[43] Besides repeating Dürer's image of the solitary rhinoceros, these two books also included a splendid, perfectly imagined scene of a confrontation between the rhinoceros and the elephant in the forests of India. The presence of the second horn and the metallic armor betray the influence of Dürer, but the antipathy between the two big pachyderms was the residue of classical natural history and was consolidated as one of the recurrent motifs in the visual and decorative arts from the Renaissance on (see Figure 7).[44] Without going through the whole list, we must mention at least one of the more original reproductions of that eternal rivalry: the magnificent series of fifty-five columns or caryatids with fauna designed by Joseph Boillot in his treatise *Nouveaux portraitz et Figures de Termes* (1592), which appeared in German

Figure 14. **Zoomorphic columns, in Joseph Boillot,** *Nouveaux portraits et figures des termes pour user en Architecture*, **1592.**

as *New Termis Buch* in 1604. In this Mannerist work in which zoology is at the service of architecture, there are two columns with an elephant (Figure 14). While the first is locked in the coils of a dragon, which it seems to be vanquishing, in the second the rhinoceros is placed above the elephant, as though restoring or recovering the victory of evil—that necessary revenge, as we saw in Chapter 2.

Finally, to celebrate the entry to Paris of Henry II of France and his wife Catherine de Medici in 1549, Jean Goujon erected a monument in which another rhinoceros copied from ours supported an obelisk surmounted by a sphere and a warrior who symbolized France.[45] The rhinoceros had already been adopted by the Medici as one of its emblems thanks to Paolo Giovio, the chronicler, collector, and physician who took such an interest in the fate of Ganda.

We will never know whether Giovio had the hide or dissected cadaver of Ganda transported from the Ligurian shore to Rome. However, it is certain that he had a copy of *Rhinocerus 1515* in his *museo*, a word that he was the first to use for the building that housed his magnificent collection of effigies and portraits of living and dead sages, poets, artists, and politicians— a collection of *imagines*—including that of our portentous oriental beast. As a major creator of emblems, Giovio was able to connect the political force of the image as propaganda with the symbolic and moral message of certain classical legends. He was the one who suggested to Duke Alessandro de Medici the use of the rhinoceros as an allegory of invincible strength (Figure 15). "I do not return without winning" *(non buelvo sin vencer)* was the Spanish motto that accompanied this first appearance of the rhinoceros in the emblems of the West, an idea inaugurated by the Medici and later picked up by others, as can be seen from the works of Camerarius and other major emblemists, some influenced by Dürer's image and others not.[46] Camerarius, for instance, preferred to follow in the wake of another rhinoceros—the one that arrived in Lisbon in 1579 and spent three years in Madrid, closer to its owner, King Philip II of Spain and Portugal.

The presence of this second exemplar, known as Bada, in the Spanish court is attested by various sources.[47] In his *Historia de las cosas más notables, ritos y costumbres del Gran Reyno de la China* (1585), Juan González de Mendoza includes the strange scene of the first Japanese embassy in Europe as it observes the

Figure 15. Emblem of Alessandro de Medici, from Paolo Giovio's *Dialogo dell'impresse militari et amorose*, 1555.

rhinoceros together with the other big pachyderm, an elephant, its inseparable sparring partner, an episode that took place in Madrid in November 1584. We know that the entry on the scene of Bada, the female successor to Ganda, triggered another wave of reproductions, some of which adopted the fabulous morphology of her predecessor. One of them even traveled as far away as the improbable location of Tunja, in the sierra of present-day Colombia, where a Dürer-style rhinoceros appears among the frescoes that decorate the ceiling of an ancestral home from the end of the sixteenth century.[48]

This is the first rhinoceros in the New World, opening another fantastic cycle, colonizing the American imaginary world, and contributing to the globalization of Dürer's engraving. The spread of the image had a temporal or retrospective dimension too: one of the classical numismatic catalogues of the eighteenth century, that of André Morell, includes some images of coins from the time of the emperor Domitian bearing the image of a rhinoceros. It must have been one of those employed in the games of 88 AD, and yet Morell's image of the Roman coin reflects an animal suspiciously like the archetype created by Dürer.[49] We can even find adaptations in Japan, China, and India of the image created by the master of Nuremberg; its universality came to shape the anatomy of the beast even in Asia.[50] The Orient itself, it seems, was vulnerable to the power of this Orientalist version and vision. It is the paradoxical return of Ganda, who travels around the globe invested with the aura and poetic inspiration of his author and of the central European Renaissance.

Among studies of the proliferation of the image of the rhinoceros in all the visual and decorative arts, including in tapestries, ceramics, and paintings, furniture, cameos, and various items of jewelry, we can single out the delightful book by T. H. Clarke, an essential resource for any essay in rhinocerology and for these pages.[51] Clarke not only traced the itinerary of Dürer's image and those derived from Bada in Madrid; he also investigated images based on two specimens that arrived in London in 1684 and 1739. These latter enabled James Parsons to outline the differences between the African and the Asian species, a study that was properly completed by Petrus Camper and Georges Cuvier before the end of the eighteenth century. Clarke also paid attention to the very rich Dutch representations from the eighteenth century and concluded with those deriving from an exemplar in Versailles and a third rhinoceros in England. It was the latter that Camper used and that was immortalized by George Stubbs, the famous painter of wildlife who is known in England above all for his portraits of horses.

In the field of natural history, the impact and hegemony of Dürer's woodcut are overwhelming. In the face of other representations and other testimonies, the imposition of the fantastic morphology of that species was persistent if not systematic. After its appearance in Gessner's zoological encyclopedia, we find it soon afterwards in the teratological tract of Conrad Lycosthenes, *Prodigiorum ac ostentorum chronicon* (1557). Ulisse Aldrovandi included an imaginative version, brilliantly hand-colored, in one of his many posthumously published works, *De quadrupedibus solidipedibus* (1639). It reappears in Jonston's *Theatrum Universale Historia Naturalis* (1660). Illustrated with splendid engraved plates, this is the last Renaissance encyclopedia, a work that connects the old-fashioned knowledge with the new winds of experimental science and modern physiology.

But neither Parsons, with his meticulous reports in the *Philosophical Transactions* and his verist drawings of the exemplar of 1739, nor the drawings of Francesco Lorenzi, nor the delightful oil painting by Pietro Longhi of the rhinoceros that arrived in Amsterdam in 1741 and was shown in Venice, nor the numerous portraits of the same exemplar that include such superb renderings as the one in *Tabulae sceleti et musculorum corporis humani* (1747), a work by the anatomist Bernhard Siegfried Albinus illustrated by Jan Wandelaar (see Figure 23), could compete with the overwhelming force of the figure created by Albrecht Dürer.[52]

The impetus of our armored creature seems to wane toward the end of the eighteenth century with Camper and Stubbs. His invincible aggressiveness is toned down—but only in appearance. The *Histoire naturelle* of the Comte de Buffon (1707–1788) and even the *Encyclopédie* are still indebted to the stereotype that was born in the Nuremberg workshop in 1515. Its survival continued beyond the Enlightenment and down into German school textbooks from the first half of the twentieth century.

As Joseph Koerner writes, Dürer is an archetype of human agency.[53] His art spreads and dominates all it touches. Through his innovative strategy of producing and reproducing images, he and his work achieved a kind of omnipresence, an undisputed sway over space and time. The artist himself is inscribed, engraved in his work through his monogram, that signature that seems to combat the natural attack to which it is condemned: piracy and indiscriminate plagiarism. Dürer was involved in legal wrangles connected with the usual problems associated with such matters, with what we would today call intellectual property. The courts protected his signature, but could do little against the copying of his images. In the era of serial art that the engraving inaugurated, authorship, originality, and copyright were delicate matters. It was not yet imitation, the copy of the copy. Falsification was the price, the natural counterpart of the ubiquity and perfection of the printing processes.

The monogram also seems to mitigate the disappearance of the author that mechanized production entails. The ubiquity that Dürer achieved by means of his engravings required an absence: his absence. It relied on an industrial, machine-made process. It has been some time now since the artist abandoned his task. The presses work far beyond the reach of the designer of the print. The work can be produced even after the death of its maker: there is no better proof of his expendability.

Mechanical reproduction, the technology that enables diffusion, also presupposes a certain rift between the author and his work. A modern distance is opened up, connected with the mediation that takes place in the world of machinery and the first dehumanization of art. It resembles those workshops filled with alembics and artifacts where the alchemists trafficked in strange substances and laid the basis of what was to become chemistry; or those observatories where telescopes and quadrants produced observation and data; or, a little later, those laboratories where scientists worked with microscopes and experimented with vacuums, light, electromagnetism, and other imponderable phenomena; the machine and technology take control in the studio and in the work of the artist to produce images, images that are and are not original. Images, like *Rhinocerus 1515*, that fashion rather than reflect the world, its phenomena, or its forms.

The history of our armored rhinoceros is a history of the force of technology and the power of an artifice like the engraving to circle the globe and domesticate perceptions. It is an episode characteristic of the first globalization. It heralds the empire of the image, but also reminds us how many words images bear with them and nourish. Instead of the fiction of the lifelike portrait and the verism of the natural representation, we feel the real weight of the shared imagination in the production of natural things, of our hopes, fears, and desires. And of the creative act, as social as it is individual (the entire sociology of art and science are not able to quench the little fire that animates what is singular and cannot be repeated). It is at all events an act that is productive inasmuch as it is reproductive, more creative than imitative, as narrative as it is mimetic. Dürer himself said that the object of art is to produce difference, not resemblance. His method was transformation; his objective was to create new, different things. Great art exists in that small space between the slavish copy, condemned to disappoint, and excessive difference, which contradicts nature.[54] We can apply this consideration to science as a practice. Only by composing fantastic images, a dream work, can anomaly be liberated and art and science generated.

So let us consider the strength of the imagination and the creative genius, but also the material side of the process, its mundane character. Instead of the false notion that things are universalized because they are true, we are obliged to accept once more the stubborn reality of the opposite: it is only through their circulation and reproduction all over the globe that they become unshakable and invincible.

PART TWO

A Strange Cadaver

CHAPTER FOUR

Chimera

There is something more bold and masterly
in the rough careless Strokes of Nature,
than in the nice Touches and Embellishments of Art.

Joseph Addison, *The Pleasures of the Imagination*

Our second protagonist crossed the ocean packed in seven
crates. Like Ganda, he did so in the hold of a ship, and he
too rested his bones on the Iberian Peninsula. He disembarked
not in Lisbon, but in the Spanish port of La Coruña. Without
leaving the crates, he then traveled overland to Madrid. His
final destination was the Royal Cabinet of Natural History,
where his majestic effigy can still be seen today.

Like Ganda, he was a large vertebrate, or at least he appeared
to be. Although there certainly are many resemblances between them—and this essay is a hazardous attempt to trace
and find them, almost to create them, to examine analogies
and parallels that, far from being evident, emerge from experimental artifice or historiographical argument—their differences
are profound and at times even strikingly obvious.

What were these differences? What separated them? The
dates, for a start. Our second animal was transported to Madrid
in 1788, 273 years after the journey of the rhinoceros, on the eve

of the French Revolution and after the completion of the reconnaissance of the globe. The West had managed to circumnavigate the globe and had started to dominate it. The second Age of Discovery was concluding the task begun in the dawn of the early modern era, the age of Ganda's peregrinations. The promise of unification held by the globe of Nuremberg, that prophetic artifact, had ceased to be a dream for adventurers, bankers, and businessmen. It had become a reality.

The creatures also had different places of origin. While Ganda came from the East, the fantastical India of legend, this second animal came from the opposite extreme of the planet, that westerly shift of the Occident's *plus ultra* that we call the New World.

It was there, on the border of the youngest and southernmost viceroyalty of the Spanish Empire, on the banks of the River Luján, some thirteen leagues from Buenos Aires, that the fossilized remains of our second protagonist appeared. It was a mysterious creature whose identity would not be revealed until later, one that would rapidly provoke intense debates in which, as we shall see, science and religion, politics and symbols, patriotism and the emergence of time and life—in short, the seen and the unseen—were to lock in combat.

But let us not get ahead of ourselves. What is the most striking, most obvious difference from Ganda? His state, of course. While the pachyderm had reached Lisbon alive and complete, our second exemplar came in pieces and was dead on arrival in the Iberian Peninsula. That is how they had found him on the banks of the River Luján: buried, dismembered, and dead, long dead. How long had he been there? Much longer than it was possible to imagine. And even though it was important to know how long he had been lying there—in fact, far more important than anyone could imagine at the time—there was another, perhaps more pressing matter to settle first. Who was the deceased? What animal possessed such enormous bones? Was it a giant?

All was doubt and speculation at first. There were only a few facts to go on, many of those contradictory, and there were many questions. This case, a classic in the annals of vertebrate paleontology, began as an insoluble puzzle, a killer Sudoku of forensic anatomy.

IT ALL STARTED in that remote gully, an alluvial terrain that favored the conservation of fossil remains. It was early in 1787 when Francisco Aparicio, mayor of Luján, received the news that some enormous bones were surfacing nearby, within his jurisdiction. He decided to inform Manuel de Torres, a Dominican friar who gave evening classes in a school belonging to his religious order in Buenos Aires.[1]

Torres was not a scientist (there were still no scientists as such at the time), but an educated local with an inquiring mind and some knowledge of natural history. He was about forty-seven years old and came from Luján, so he was familiar with the place, and in fact used to conduct excursions in the hope of finding curiosities, botanical and zoological species, minerals, and even fossils.

Torres got down to work and ended up disinterring the skeleton. He had to get over several hurdles. The technique of exhumation was not well developed at the time (these were the early years of paleontology). And then there were the usual problems involved in complex works of discovery, the sorts of episodes common in the history of science: corporate disputes, rivalries about who was first to do what, and other such niceties (the fact that there were not yet any scientists in the sense of autonomous or institutionalized professionals does not mean that their forerunners were exempt from the vices of the profession).

We know about these activities from the correspondence that Torres maintained with the mayor, Francisco Aparicio, and with the viceroy himself, Nicolás del Campo, the Marquis of Loreto, who took a great interest in the discovery beside the River Luján from the very first.[2] The discovery suited both the spirit of Bourbon reformism and the scientific policy of the secretary of state of the Indies. Moreover, it provided a response to the explicit request to collect natural products and send them to the Royal Cabinet of Natural History in Madrid, an institution that had been in existence for barely a decade and was very much in need of pieces like these. Loreto rapidly sized up the situation: that strange skeleton offered a good opportunity for the young viceroyalty to distinguish itself in service of a scientific policy that was experiencing its heyday.

On April 29, 1787, Torres was delighted to be able to report to the viceroy that he had removed all the soil from above and around the bones and now had a skeleton of the whole animal. He added that he did not dare to move it before a draftsman had been sent to put it down on paper to avoid spoiling the work and depriving the world of the pleasure of being able to see something quite extraordinary. The draftsman was to draw the skeleton before its removal to preserve its form and to facilitate its reassembly after removal, in order that "this marvelous and providential work of the Lord be made known to the public." The Marquis of Loreto granted this request on the next day and sent Francisco Javier Pizarro, an artillery lieutenant, to make a "precise drawing before it is moved and risks the dislocation or fracture of its parts, also recording the dimensions of the whole and of the separate parts."[3]

However, the collaboration between the friar and the military engineer became a battle of reciprocal accusations regarding their respective roles. On May 9, Torres informed the mayor that the bones had been considerably damaged and asked him to place a guard on duty night and day to avoid any further attacks. Pizarro claimed that Torres was incompetent and had supervised the exhumation in a sloppy manner. In his opinion, the friar was a dimwit and quite incapable of supervising such an operation. Torres counterattacked and accused his opponent of impugning his honor. He complained to Loreto about the insults he had received and declared that not he but the engineer had damaged the skull and hip of the strange skeleton; the connection between the skull and the spine had been damaged beyond repair, and the hip had been broken during the operation of raising it for mounting.

Whatever the merits of the case, the bones were removed, crated, and taken to Buenos Aires, where they remained from July 1787 to March 1788. It is not known whether Pizarro managed to draw the skeleton on the banks of the River Luján, but once it was in Buenos Aires, it was drawn by José Custodio Sáa y Faria, an official cartographer of Portuguese origin in the service of the Crown and a person of some standing in the scientific and political life of the viceroyalty.[4] We do not know if he was the first to draw the bones, or whether he followed Pizarro or Torres, but at any rate he made or copied two drawings that accompanied the skeleton on its voyage to Spain.

They still survive today.[5] The two pen drawings with gray watercolor shading are a very important graphic testimony in

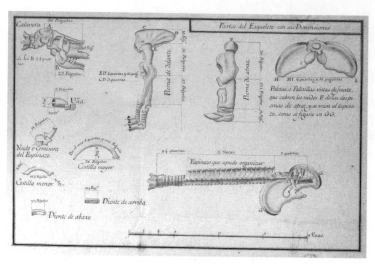

Figure 16. José Custodio Sáa y Faria, Partes del esqueleto con sus dimensiones (Archivo General de Indias, Seville).

this whole history. One of them, as its caption states, represents "parts of the skeleton with their dimensions" (Figure 16). It is an arrangement of some of the animal's main bones: the skull, a claw, one front and one rear leg, the spinal column, two teeth (one each from the upper and lower mandible), two ribs of different sizes, and the hips. They are separated from one another as in an anatomical atlas in which each bone is given independent treatment.

The other drawing, however, is the result of the first reconstruction of the skeleton, though we do not know whether such a physical reconstruction actually took place. While Viceroy Loreto refers to this drawing in an official note as "the figure that seems to match [the skeleton] or that would have if it were put together," it is possible that the skeleton was assembled for the first time in Buenos Aires and that the drawing reflects that

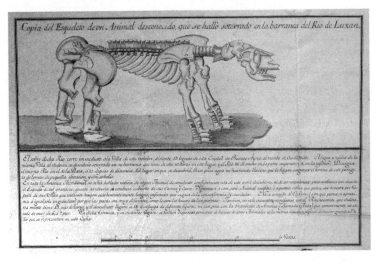

Figure 17. José Custodio Sáa y Faria, *Copia del esqueleto de un Animal desconocido que se halló soterrado en la barranca del Río de Luxán.*

reconstruction.[6] We cannot be certain what really happened. If Sáa y Faria imagined his reconstruction and the first reassembly of the cadaver from the River Luján was virtual, that would make the episode all the more interesting, since however fantastic and imaginative the drawing may be, it nevertheless inspired and guided the reconstruction of the animal months later in the Royal Cabinet in Madrid. The caption on this drawing is significant: "Copy of the skeleton of an unknown animal that was found buried on the shores of the River Luján" (Figure 17). Is this copy a falsification, an alteration of the original? If any image or copy is a re-creation and distortion of reality, this is such a case.

The bones had been found together, but they were separated and scattered. The first task facing the discoverers of the animal—and later all those who studied it—was to assemble it. Those *disjecta membra* had to be reconnected to give it a form. It was like a jigsaw puzzle without the most important and essential part: a model to follow. The basic question was how to assemble the skeleton of an unknown animal. What shape were they to give a creature that was not just strange and mysterious, but completely without precedent? Nobody had ever seen anything like it before. The *copy* to which Sáa y Faria referred was bound to be an innovative work, more creative than reproductive, more poetic than mimetic. Without close reference points, and given the contradictory indications of its identity, it required a good deal of imagination. To compose an image of the reassembled skeleton in order to see the animal, the first task was to imagine it, to give it an image.

After all, what kind of an animal was it? Who had been unearthed on the banks of the River Luján? The question was met with different answers, and it took some years before a satisfactory, surprising, and revolutionary one presented itself. At the time, what had arrived in Buenos Aires was a bundle of unconnected bones. They included eighteen vertebrae to be arranged to form a spinal column, from the hip to the head, some three and a half meters long (four *varas*). If we add a head more than thirty inches (almost 70 centimeters) long, we end up with a creature some four and a half meters long, including the neck but excluding the legs. The sacrum alone weighed around 175 kilograms.[7]

And that was not all. The head was impressive not only for its size, but also for its peculiar shape. Beneath the very elongated skull were the jawbones in which some broad teeth shaped like molars were fitted. If we add that, attached to the extremely robust legs were feet with toes ending in a kind of claw with what appeared to be sharp nails or some other kind of sheath, the whole was more than just extraordinary; it was prodigious. An herbivore with the claws of a carnivore? A feline the size of a pachyderm?

At first sight it was a chimera, a creature combining the parts or attributes of other animals, and this is a view that was to dog it for many years afterwards. This is apparently what Sáa y Faria made of it, to judge from the long description beneath his "Copy of the skeleton," recalling the one that accompanied Dürer's rhinoceros. Extraordinary creatures require explanation, a discourse, or an incantation in an attempt to explain the baffling. Impossible images call for words. The second paragraph of this description runs:

> In the whole of South America nothing is known of any animal shaped like this one now discovered nor of its bulk; for given the size of the skeleton, how much bulkier must it have been when covered with flesh or hide. It is not known whether it was an amphibian or aquatic animal, although it seems to be terrestrial to judge from its claws, which must have been quite long, as may be inferred from the circumference of its toes. It does not resemble the elephant (to which it approximates in size) because its feet are very different, as are the femurs. Nor does it resemble the rhinoceros, which is usually 13 feet long, while the one discovered was 18 and had a different shape. Neither does it resemble the great beast of America (called Anta) which commonly does not exceed 6 or 7 feet. Pieces of the bones of other animals of the same species and some smaller than what this copy represents were found in this gully and in various other places.[8]

Above these words was set the very rigid image of an animal in the position of a quadruped resting, with its front and rear legs completely straight and the spinal column parallel to the ground (see Figure 17). It is evident what the points of reference were when he was given his first identity soon after being unearthed. Whoever reassembled him or drew a reconstruction of him—Torres, Pizarro, Sáa y Faria, perhaps other advisors to the viceroy—had in mind either a pachyderm or an enormous feline. The explanatory comments by Sáa y Faria reflect the uncertainty: it looked like a terrestrial animal, although the possibility of its being aquatic or even amphibian could not be ruled out. It was unlike an elephant or a rhinoceros (unfortunately—the hypothesis would have provided a curious variant for our argument). Nor was it a tapir (anta). Unlike Ganda, this animal was not hoofed; instead, it had claws like a carnivore. But who had ever seen a carnivore five meters long and more than two meters tall? And if it was a huge wild cat, a gigantic kind of tiger, why did it have molars for teeth? That fierce creature, to judge from its claws, did not have incisors or tusks of any kind; its teeth were quadrangular prisms like those of an herbivore.

IN GREEK MYTHOLOGY the Chimera was a monster with a lion's head, a goat's body, and a serpent's or dragon's tail. In other versions it had three heads, each corresponding to a part of its strange anatomy. This is how it appears in the famous bronze from Arezzo (Figure 18). Its behavior matched its appearance: the Chimera was a terrifying creature that devoured animals and humans.[9]

Almost every culture has had its own chimeras, its imaginary zoology of the fantastic. Jewish tradition, for example, had its

Figure 18. Chimera of Arezzo, Etruscan bronze *(above)* (Museum of Archeology, Florence), and a reproduction of it in a public fountain in Arezzo *(below)*.

composite creatures, ancestral China its lions or hounds of Fu, while the Aztecs had their cult of Quetzalcoatl, the plumed serpent. Griffins, centaurs, and sphinxes are all the result of strange, imagined combinations, but it is the Chimera that has always represented the quintessence of hybridization to such an extent that its name has come to stand for something unreal, imaginary, impossible. In fact, the Dictionary of the Royal Spanish Academy also defines the word as "that which is presented to the imagination as something possible or real, without being so." A chimera is an unreal fantasy, an impossible aspiration, an empty belief. But it has other meanings too.

In the field of genetics, a chimera is an organism composed of two kinds of cells of distinct origin—that is, from different zygotes. This is occasionally found in nature, but genetic modification has changed all that. A chimera is the result of a teratological experiment based on hybridization between different species. In this sense it could be said that, instead of burying chimeras, science has made them possible. Finally, the term is also found in paleontology today, where it is used to designate fossils composed of parts of individuals of different species that were believed to belong to the same species at the time they were discovered.

There is something of all this in our story. The teeth and claws of the creature from the River Luján seemed to come from different animals, but had been exhumed in the same spot without any indication of the presence of two individuals (the remains included only one skull, four legs, and one spinal column). The paradox was that the teeth and the claws were incompatible but could not belong to two individuals of different species. This made an even more remote possibility attractive: that they were the remains of a giant, an idea that at any rate was soon dissipated.

The confusion was very common. Ever since antiquity the bones of extinct large vertebrates had frequently been interpreted as belonging to gigantic human beings of the past. America as a whole had been fertile ground for hypotheses and legends of this kind. The indigenous cultures themselves generally attributed the large bones to the giants of the past, if not to fabulous (chimerical) animals. In New Spain, for example, the Toltecs believed in an extinct race of giants (*quinametzin*), and the Europeans who arrived in the colonial period did not lag behind. Almost all the chroniclers—Gonzalo Fernández de Oviedo, Francisco López de Gomara, Pedro Cieza de León, Francisco Hernández, José de Acosta, Tomás de Torquemada— interpreted the large fossil remains as the remnants of giants. And as if that were not enough, the Río de la Plata was not so far (in relative terms) from Patagonia, one of the epicenters of the universal belief in giants, where more testimonies and evidence were accumulated. The navigators and travelers of the eighteenth century (John Byron, Antoine-Joseph Pernety, Thomas Falkner) took it upon themselves to revive the myth initiated by Antonio Pigafetta, who had circled the globe with Magellan in the sixteenth century. As a result, the theory of Patagonian giants had one of its most glorious moments in the heyday of the Enlightenment and reached the dizzying heights of the hypothesis of a race of giants who had lived before Adam and the Flood. The major contributor to this was undoubtedly the Franciscan José Torrubia, whose *Gigantologia spagnola vendicata* (1760) was a vast compilation and ordering of arguments of every kind (ancient and modern, anatomical and philological) to demonstrate that the large fossil remains discovered in the New World had actually belonged to an extinct race of giants.[10] In the second half of the eighteenth century some Jesuits and engineers made discoveries in the Río de la Plata that, in one of those anachronisms common among historians, have been called the first stage of "Argentine paleontology," even though neither Argentina nor paleontology existed at the time. A century later, by which time they both did exist, collectors and explorers heard the local inhabitants say: "If you are looking for fossils, all you have to do is ask for giants."[11]

However, once the skeleton had been extracted from the gully of the River Luján, it was clear that it could not have belonged to a human being, no matter how large. Those were not the bones of a giant, and since they did not belong to two individuals of different species, they could not be regarded as a chimera in the paleontological sense of the word either. And yet, the mysterious animal was both gigantic and chimerical thanks to its two main characteristics: its extraordinary size and its peculiar morphology, or as Sáa y Faria put it, its bulk and its shape. It was a creature of colossal dimensions and with a composite anatomy, in other words, a *gigantic chimera*. Its nature bordered on that of a prodigy or a portent.

Was it a monster? In a certain sense it was, which explains why its finder, Manuel de Torres, described it, as we have seen, as "this marvelous and providential work of the Lord." In fact, it was a monster in the sense of an anomaly, a departure from the norm, an irregularity of nature, and therefore a singular and exceptional phenomenon capable of arousing both astonishment and great scientific interest at the same time.

Katherine Park and Lorraine Daston have demonstrated how the language of marvels and prodigies, including teratology, continued to play an important part at a time when modern science was fully under way.[12] The study of monsters, rarities, and other curiosities had become a secular activity and acquired respectability and dignity in the Renaissance courts, where they were the most sought after items to populate the *Kunst- und Wunderkammern*. Although their zenith coincided with Mannerist taste and a Baroque passion for the strange and deformed (the period in which treatises on teratology flourished can be fixed between 1580 and 1680), neither empiricism nor rationalism managed to eradicate them completely from the intellectual panorama of the eighteenth century.

Certainly, the *philosophes* threw many of the prodigies, monsters, and marvels onto the bonfire of superstitions and popular beliefs. The disenchantment of the world produced by modern mechanistic theory and rationalism and the progress in physical and natural science considerably reduced the number of phenomena that could still be categorized under those headings. During the Enlightenment the Republic of Letters managed to impose a taste and natural order that prioritized simplicity, mathematical laws, and harmony.[13] All the same, a few monsters and chimeras still managed to survive the Enlightenment onslaught. After all, the naturalists, physicians, and philosophers of the modern era were interested in monsters because they were extreme cases of singularity. They were isolated and exceptional phenomena that the new scientific gaze ought to account for, unlike the scholastic precepts that recommended obtaining knowledge from the regularities and ordinary course of nature. A monster or any other extremely singular natural phenomenon such as an eclipse, a comet, or an exotic plant, attracted the same scientific interest as those other singular phenomena that were produced artificially in laboratories by the first experimental scientists. Both types of phenomena occupied the marginal zone of nature, which was shadowy but promising from the point of view of knowledge. A two-headed calf, a malformed colt, a bearded woman, a giant, or a hermaphrodite furnished opportunities for reflection and astonishment. They mobilized the passion of the soul that from Aristotle to Descartes triggered knowledge: the sense of wonder.[14]

So when our skeleton from the River Luján made its appearance, interest in the extraordinary had not died out completely. In fact, if there was an ideal ground for the emergence of the penultimate natural prodigies, this was the New World, a continent that had come onto the scene during the Renaissance decked out with the rhetoric of the marvelous and where there was still room for surprises, monsters, and chimeras in the Enlightenment. Through one of those fabulous translations with which Europeans fabricated our idea of the world, America had inherited many of its attributes from the Orient. Not only biblical narratives and geographical myths, but also the fabulous zoology that had once been the exclusive property of the East, had been projected onto America. The moral and natural histories of America had paraded a splendid catalogue of zoological and anthropological rarities: birds without feet, dog-headed men, armadillos, Amazons, giants, beardless lions, huge reptiles, and other extraordinary creatures.[15] Through its remoteness and its diversity, American nature was monstrous itself, among other things because of the inability of the language of the first *conquistadores* to describe it. To describe certain species in a recognizable way, they had to resort to aggregates of the fauna they knew. Thus the American bats were flying cats, the Patagonian *Su* was a cross between a squirrel and a lion, and so on. For many reasons of various kinds, America was a continent that was prone to producing the marvelous, the chimerical, the monstrous, and even the diabolical.[16] Our fossilized skeleton, in this sense, was a creature as extraordinary and prodigious as the rhinoceros had been in the sixteenth century. That too, as the reader will recall, was described by borrowing terms used for other, better-known animals (the elephant, the bull, the wild boar). In this respect they were similar. Both the rhinoceros and the fossil were extraordinary, surprising animals. Their heterogeneous, chimerical morphology turned them into fabulous creatures. They were, or had been in their day, real animals, but they could have been taken from the pages of a compilation of the fantastic.

In the classical tradition monsters were interpreted as signs (*terata*) of the gods or of divine providence; they were divine messages, omens of future events, phenomena with a certain miraculous character. A monster was not just something that went against the regular order of nature—"the background noise, the endless murmur of nature";[17] it was also a warning or a manifestation because it revealed the great and awful events of the future, such as the death of a monarch, a plague, or a victory in battle. But what did that monster dug up on the banks of the River Luján in 1787 reveal, that gigantic chimera composed of fearful claws and an herbivore's teeth? What extraordinary event did it prophesy?

Unlike the classical monsters, this one did not descend from the skies. It had emerged from the bowels of the earth. Rather than foretelling the future, it had a message about the past. It did not seem to be transmitting a sign from the gods, even though, as we shall see, there were those who wanted to interpret it as evidence of Providence. Those fossil remains bore the signs of something equally invisible, more mysterious than all the angels and the courts of heaven, or, to put it differently, less familiar at the time, a phenomenon that may seem common enough to us today but was completely unforeseeable then: time, the history of life and of the earth.

Nonetheless, nobody could have been aware of this at the moment of its discovery. Few naturalists adhered to the theory of extinction in the eighteenth century. The species were fixed; only a few considered the possibility that there had been other, different ones in the past. In a world in which there were still large tracts of land to be found and explored, the discovery of the remains of an unknown exemplar did not imply that it was an extinct animal. Perhaps one fine day a traveling naturalist would bump into another, living exemplar of the same species in some corner of the planet.

In fact, once the skeleton had arrived in Buenos Aires, the Marquis of Loreto organized a meeting of several of the local representatives of the *indios* of the region *(caciques)* to ask them about the cadaver. It may seem paradoxical to us, but while the enlightened viceroy wanted to know whether they recognized the animal and whether any live specimen might be found thereabouts, apparently the *indios* soon abandoned their belief in giants and opted for the hypothesis of an extinct species. Faced with the strange case of the River Luján, the knowledge of both sides was equalized; the "civilized" were just as disorientated as the locals—or more so:

> Several *caciques* of the natives *(infieles)* of the Pampa and the Sierra recently came to the capital; I took care to show them these bones and the way in which they had been arranged to complete the figure of this animal. They were surprised and assured me afterwards that they

could not be from this area because of the absence of any
report of them; they had always thought that they were
the huge bones of one of their ancestors, but it would be
quite normal if they had wiped out these animals, if they
were dangerous and not very common, when they were
the only lords of these lands.[18]

Once the hypothesis that the bones belonged to a gigantic
human had been rejected right at the start, the theory that they
came from an exemplar of some unknown living animal, some
rare species that lived in the Pampas or in Patagonia, raised certain expectations. On September 2, 1788, the minister Antonio
Porlier conveyed to Viceroy Loreto the wishes of His Majesty
to ascertain, by all the means at his disposal:

> whether a living specimen of the same species as this
> skeleton, however small, could be found in some part of
> Luján or somewhere else in the viceroyalty, and to dis-
> patch it alive if possible, and otherwise dissected and
> stuffed with straw, duly assembled in a lifelike way, with
> all the other necessary precautions, in order that it should
> arrive in good condition and that His Majesty may have
> the pleasure of seeing it whenever he chooses.[19]

HOWEVER, NEITHER THE king nor any of his subjects had
their curiosity satisfied. It proved impossible to capture one
alive, let alone to stuff it. The last exemplars of that species
had died thousands of years earlier. The authorities had to
make do with these fossilized and disjointed remains from
the River Luján, which were dispatched from Buenos Aires in

the spring of 1788 and reached El Ferrol on board the packet frigate *Cantabria* in June or July of the same year. In his letter of September 2, Porlier acknowledged receipt of the seven crates with "the skeleton of the animal unknown to the naturalists." It was sent by land from El Ferrol to the Universal Office of the Indies in Madrid and transferred soon afterwards to the Royal Cabinet of Natural History "for the skeleton to be assembled and identified by the experts in natural history."

That is how the wonder of the Río de la Plata reached the establishment in Madrid on the second floor of the palace of Goyeneche in the Calle de Alcalá, at the time the seat of the Academy of Fine Arts of San Fernando, which now occupies the whole building. It was the wish of Charles III that "the sciences and the arts, associated for the public good, should come together under one roof" in this building, as the inscription engraved on its façade proudly stated when it was opened to the public in 1776.[20] The Royal Cabinet of Natural History, the predecessor of what is today the Museum of Natural Sciences of Madrid, was the result of a long institutional search, a space created to fill a notable gap in the metropolis of what was still the largest colonial empire in the world. As one of the few European countries that lacked an academy of sciences, Spain could finally count at least on a cabinet of natural history to house and exhibit the products and creatures of its American and Oceanic provinces. In 1771 it had acquired the private collection of Pedro Franco Dávila, a native of Guayaquil who lived in Paris. This collection formed the nucleus of the fledgling cabinet, with Dávila himself serving as its first director. Although the early years of the institution were checkered ones as a result of the lack of resources and also, certainly, of expertise, it is fair to state that some achievements were made, of which the most important was probably the founding of an institution that has now been in existence for almost two and a half centuries.

In fact, Dávila's collection was one of the best private collections in Enlightenment Paris, where the distinguished Creole had formed it in the course of "more than twenty years of dreams and investigations."[21] It was especially well provided with minerals, shells, marine products, and petrified objects, that is, fossils and curiously shaped stones. It also included a good number of manmade objects: works of art, ethnographic items, furniture, instruments, paintings by Bartolomé Murillo and Hieronymus Bosch, and even a Velázquez. Once the collection had been installed in Madrid, it was augmented with other items from the royal collections (the old House of Geography, the treasure of the Dauphin, and others). But above all instructions were drawn up and committees were created to collect products and creatures from Spain, the Americas, and the Philippines. Objects, minerals, animals, and a plethora of curiosities were not slow to arrive from both shores of the Atlantic and the Pacific. The successive overseas scientific expeditions and similar scientific commissions on the peninsula generated such a volume of shipments that the vice director of the establishment, Clavijo y Fajardo, ended up calling for a halt to further dispatches in 1791. In his view, the Royal Cabinet now ran the risk of looking like a "funfair."[22]

Was it? Not at all, although the comment expresses some significant fractures: notably, the gap between the capacity of the Crown to collect and its capacity to store and display those collections in a space with the proper infrastructure and scientific staff; and the disparity between two forms of collecting, one that was close to the Baroque *Kunst- und Wunderkammern* centered on the culture of the prodigious and the marvelous, and one that was more rigorous and bound by scientific criteria, in which scholarly naturalists, physicians, physiologists, anatomists, and geologists held sway.

This was, roughly speaking, the profile of the establishment where the bones from the River Luján arrived in the summer of 1788. Within its walls there still existed a somewhat archaic air that contrasted with the talk of reform and modernization with which it had been created. A review of the shipments that arrived at the time indicates a fascination with what was rare or hybrid. In September 1787, for example, it received a *danta* or "large beast" (a tapir). A month later Flordiablanca (1728–1808) sent Dávila's successor, Eugenio Izquierdo, "a stone from which emerged an olive plant half a meter tall and a monster with a body like a watch called a *zaratán*" (a sea turtle). In the following year a report arrived about the existence of "a very fierce species of satyr, half human with long white hair and half horse with big round hooves, which lived near Popayán and the mountains of the plain of Guanacas or Puerto Nevado, in the viceroyalty of Santa Fe."[23] Later arrivals included Siamese twins, a one-eyed dog with a trunk, and a monstrous sheep with eight feet and a single head with four ears. Tapirs, giant sea turtles, satyrs, Siamese twins, and other deformed creatures: these are the imaginary or real creatures that demonstrate the teratological character of the Royal Cabinet when the Luján skeleton arrived. Inevitably, it was not long before it was labeled "the monstrous skeleton," "the big monster from the River Luján," and suchlike.[24]

It is not a coincidence that the title of the first publication promoted by the Royal Cabinet is symptomatic: "Collection of plates representing the animals and monsters of the Royal Cabinet of Natural History in Madrid (1784–1786)." It was the work of Juan Bautista Bru de Ramón (1742–1799), the painter and taxidermist of the Royal Cabinet who was directly responsible for the assembly and first description of the skeleton there. This controversial figure, who was to play a peculiar but decisive role in this story, was the object of various polemics in his day, though exaggerated claims in his defense have been made by the well-known historian of Iberian science López Piñero.[25]

Bru came from a well-to-do family in Valencia. He may have trained as a painter in the Academy of St. Barbara, and certainly studied under Francisco Bayeu (the tutor of Goya). He executed the frescoes in the church of the Madonna of the Rosary in Canyamelar, Valencia, but after his arrival in Madrid in 1766 he came to specialize in anatomical painting and natural history. From 1773 he worked occasionally as a taxidermist for the Royal Cabinet and began permanent employment as the "first taxidermist" four years later. From then on his work was connected with the painting, preparation, taxidermy, and assembly of animals, especially vertebrates, in the museum in Madrid.

Starting in September 1788 and continuing for the next two or three years, he supervised the assembly of the skeleton and produced a fairly detailed description of it as well as a series of twenty-two drawings on which five high-quality, large-format plates were based. These were engraved by Manuel Navarro, a skilled engraver, and included in the publication that was promoted in the following decade by an engineer and associate of Bru, Joseph Garriga. The completed work, "Description of the skeleton of a very bulky and rare quadruped, which is kept in the Royal Cabinet of Natural History in Madrid," was published in Madrid, in the print shop of the widow of Joaquín Ibarra, in

1796, and featured the plates accompanied by Bru's own written description.[26]

There is no need to put it delicately: Juan Bautista Bru was not a naturalist in any sense of the word; he was a painter and taxidermist of dubious repute. His work on the skeleton from the River Luján had the merit of rescuing the bones from obscurity and publicizing them throughout Europe and further afield, but he was not the right person to perform a task of this nature. His knowledge of zoology was poor, let alone his knowledge of vertebrate paleontology. In this sense, his defects reveal a more serious one: that of an institution that did not have qualified personnel to carry out this task. Bru was given a job that was clearly beyond his capabilities, although, in fairness it would probably have been too much for most, perhaps all, of the naturalists, anatomists, zoological experts, and fossil collectors of the time. It would take a towering genius to solve the mystery: Georges Cuvier.

Be that as it may, it is useful to indicate the three tasks that Bru did carry out: the assembly of the skeleton, its description, and the creation of the drawings on which the plates were based. Regarding the first of these, the accounts of the Museum of Natural Sciences and the opinions and testimony of several later naturalists show that he committed several mistakes and made some rash judgments regarding the reconstruction. The list of materials used to assemble it speaks for itself: wire, polish, pitch, glue, white powder, cork, a hand saw, and files. A locksmith was paid a considerable sum of money for work connected with putting together the skeleton, and a carpenter appears on the payroll too.[27] Moreover, "a mule's skull was purchased to give an idea of how to arrange the skeleton's skull, and the tail was purchased at the same time."[28] Faced with the lack of a living or dead exemplar to serve as a model for the arrangement of those bones, Juan Bautista Bru settled on the points of reference that were closest at hand (and even prosaic, as the case of the mule shows). The mysterious animal must have been a big cat or, more probably, an herbivore. Our bold taxidermist sawed, filed, and cut various bones, filled many others with cork, placed several of them in the wrong place, added others that did not even exist, and in general altered the anatomy of the big vertebrate to such an extent that he gave it an incorrect posture—or, one might say, a "very bad" one—the epithet used years later by Mariano de la Paz Graells, director of the museum and an important Spanish naturalist during the reign of Isabel II (1833–1868), when he enumerated the many errors committed in assembling the skeleton.[29]

The monstrous skeleton from the River Luján was put together as if it were a mule or a horse. With some minor changes, this is the position in which it can still be seen today, a pose quite unlike reconstructions of other exemplars of its species in other museums, but interesting to the historian of science because it shows great respect for the scientific ideas of the past (Figure 19).[30] Placed on a magnificent pedestal, the new resident in the Royal Cabinet soon became its most famous personality. Standing on four legs, with its pelvis in a rigid and forced position, that animal would captivate the interest of the experts and the imagination of the public at large thanks to its two most striking attributes: its colossal size and its extraordinary shape.

It also continued to puzzle discoverers and natural historians. The controversies that Torres and Pizarro had inaugurated during its exhumation in Luján were to continue after its arrival in Madrid. Let us examine the confused story of what

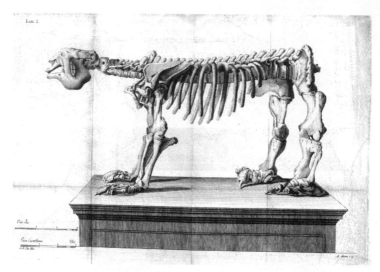

Figure 19. Juan Bautista Bru and Manuel Navarro, plate number 1, published in José Garriga, *Descripción del esqueleto de un cuadrúpedo muy corpulento y raro, que se conserva en el Real gabinete de Historia Natural de Madrid* (1796).

happened with Bru's description of the skeleton, his drawings, and the plates engraved by Manuel Navarro. It has been thought that they were produced between 1789 and 1793 and that the description and the plates did not go into print until Joseph Garriga published them in 1796. By then the news had spread around Europe and Cuvier had already published his first article on the strange animal.

If, instead of believing Garriga and Bru, however, we listen to the version of the story told by Clavijo y Fajardo, the vice director of the Royal Cabinet, the truth appears very different. In 1796, during one of the numerous disputes in which Bru and Clavijo were embroiled, the latter informed Godoy about an alleged intrigue involving Garriga and Bru, who had "recently sold as his work the description of the big skeleton in the Royal Cabinet that was made by the first surgeon of the Royal Hospital of Buen Suceso, who has come to complain bitterly to me about this plagiarism by Bru."[31]

It is a delicate matter. Although this was not the only occasion on which Bru was accused of plagiarism, we cannot prove it in this case. What is beyond doubt is that the "Description of the skeleton" consists of two very different parts: one general, the other more specific. The latter is a piece of osteological literature that shows a certain familiarity with vertebrate anatomy (the pieces are designated and described with the terminology of human osteology, which suggests that it was not written directly by the surgeon, but perhaps by our taxidermist with the assistance of the surgeon or some other expert). The first part, in contrast, is a more literary and speculative text that does indeed seem to have been written by Bru himself and is in any case interesting for our argument since it reveals the aesthetic and emotional impact produced at the time by the appearance of the portentous skeleton.

According to Bru, its sight was "one of the most attractive, promising and agreeable spectacles that can be imagined."[32] Its bulk and the enormous volume of its bones in their entirety were "astonishing and admirable." It was a "vast mass," a "rare and singular prodigy." Its head displayed an "unspeakable monstrosity," even "a certain air of displeasing and unattractive disaffection," although at the same time "an elevated eminence." Its powerful jawbones exposed enormous disfigured and eroded cavities "that silently showed what they had been." Its paws and feet were "strange, rare, and prodigious"; they were enormous masses, but adapted "for those important needs and uses that may be inferred." What needs were these? Although he inclined toward the hypothesis of an herbivore in assembling the bones, in his description Bru settled for a grandiloquent and sensational one more suited to his fiery rhetoric: "Perhaps that breast housed all the anger of the elephants, all the fury of the lions, and all the raging of the tigers."

Clearly our anatomist had in mind a fierce predator, which justifies the mention of "forefeet armed with hooks that, once animated and set in motion, would be capable of tearing and ripping to pieces the hardest oaks." It was a unique natural product, a demonstration of the wisdom of Nature. Carried away by his imagination, though not lacking in intuition, Bru speculated on the possibility that Mother Nature had arranged and imparted:

> new and unprecedented shapes and directions to a quan-
> tity of material, which in the course of time might have
> ended up as a leopard or other creature through other
> new and different mutations and changes from those that
> it has undergone so far, and does not hide the fact that at
> some time she gave thought to the incessant variety and
> mutability of forms.

He was not far off the mark here. Although he got the species wrong, he was correct in referring to the mysteries of the great chain of being, too soon to be disclosed, enveloped as they were at the time in a murky current that was a mixture (like Bru's statement itself) of the principle of plenitude and a speculative transformism, a current that was visible in the Enlightenment, although it was very different from what we understand today by evolutionism.

But leaving aside for a moment the question of what his words anticipate or not, or of what is original in them and what is not, they throw light on a cultural dimension of a scientific event: the impact produced by the contemplation of the skeleton from the River Luján once it had been assembled and displayed to the public on its pedestal in the Royal Cabinet. In these rhetorical arguments taken from the culture of the marvelous and the prodigious—the weight of the tradition of teratological literature is considerable—we can also detect the influence of the terrifying sublime, an aesthetic that had already become established at that time in revolutionary Europe. Besides the classical Enlightenment references, Edmund Burke and Immanuel Kant, some of its antecedents can also be found in such works as the *Sacred Theory of the Earth* by the theologian Thomas Burnet.[33]

Contemplation of the monstrous skeleton from the River Luján, in effect produced a certain pleasurable sensation of terror, the "Mixture of Delight in the very Disgust it gives us," as Addison called it in *The Pleasures of the Imagination* (1712). On the other hand, the object formed part of a royal collection, a cabinet that was open to the public but linked with the figure of the monarch and royal patronage. Like the Renaissance Medici or Louis XIV in Versailles, Charles III also delighted in surrounding himself with exotic animals. He received his first elephant as a gift after he became King of Naples in 1735, and allowed it to wander through the gardens of the Palazzo Reale de Portici. Later, after succeeding to the Spanish throne, he asked Mazarredo, one of the great marine experts of the period, to bring him another elephant that had arrived in Cadiz from Manila in 1773. Its destination was Aranjuez, where the monarch had a menagerie. He was known as the "big elephant" (a smaller one arrived later), and when he died in 1777 it was Juan Bautista Bru, our anatomist, who prepared the cadaver for exhibition in the Royal Cabinet. It was a remarkable piece of taxidermy. He separated the skin and the bones to produce a double reassembly of the pachyderm, as can still be seen in the Museum of Natural Sciences in Madrid. Our two stories thus have another point in common: both the rhinoceros in Lisbon and the skeleton in Madrid had been preceded by an elephant, the commonest and most popular of the large exotic quadrupeds, the natural prodigy favored by European royalty over the past two thousand years.[34] Charles III also received an anteater from Buenos Aires in 1776.[35] He kept it in another of the royal menageries, one in the Palacio del Buen Retiro in Madrid, which had been operative since the time of the Count Duke of Olivares in the seventeenth century. At the end of the eighteenth century, surrounding oneself with exotic animals was still a royal prerogative; what was rare continued to add to the prestige of the court.[36]

So before the arrival of the paleontological remains that concern us here, the Royal Cabinet was already displaying the skeleton of a large pachyderm. They were both royal possessions, wonders of nature that were displayed like works of art, natural products transformed through taxidermy and assembly into sculptures.

Destined to be contemplated and admired, the new skeleton embodied the characteristics of the sublime: grandiose, (extremely) novel and beautiful, terrifying but attractive. Bru drew a parallel between the elevations of its bones in their regular proportions and the vista of a beautiful landscape, "for it is almost like the view of a range of mountains seen from the distance on a clear and calm day when the clouds have left the horizon."[37] It was quite logical for the bones of the large vertebrate to arouse a sensation analogous to that produced by the contemplation of mountains, a precipice, ruins, or a cemetery. The response they provoked was one of silent depths, of unspeakable monstrosity. Although nobody, much less Bru, was yet able to know or calculate the age of the strange cadaver, from the first its admirers could feel the distant vistas that it opened up, the vertigo produced by being sucked into the vacuum of time.

Like any other chimera, the one from the River Luján triggered a range of emotions ranging from stupefaction to terror and veneration, a mixture of reactions that connects it to science, religion, and art. In the last resort, the monstrous skeleton of an unknown animal that had been dead for millennia had been unearthed and raised on the altars of a museum of natural history, one of those spaces that would be transformed into secular pantheons from which to spread the new gospel.

CHAPTER FIVE

Bones

And your bones shall flourish like an herb.

Isaiah 66:14

B ut let us return to the basic question: Whose bones were they? Who was the mysterious quadruped erected on the pedestal of the Royal Cabinet? If we confine ourselves to what was known at the time, we can see the depth of the mystery that contemporaries faced. It seemed to be an impossible animal, an apparently insurmountable challenge to visualize, let alone assemble. All that seemed clear was that these remains were the fossilized bones of a monster.

First, it was a gigantic and chimerical cadaver, a monstrous creature. Its large dimensions and hybrid morphology aroused curiosity, admiration, and even that nervous tremor that is sparked by the sight of something extraordinary and incomprehensible. It was a prodigious phenomenon worthy of being exhibited in a cabinet like the one in Madrid, a place midway between the old chamber of curiosities and the modern museum of natural history. Second, it was also clear that it was a fossil, a petrified organic product, a trace or vestige of nature, a paleontological remain (we shall discuss this aspect in Chapter 6). Third, and more prosaically, there could be no doubt that it was a bundle of bones, or, as some were to put it, "a vast mass of bones," "an ensemble of disjointed bones."[1]

Unlike Dürer's rhinoceros, which was so inseparable from its hide, the armor that has come to indicate its order (that of the pachyderms, the thick-skinned), our second protagonist was reduced to nothing but bones. Each had what the other lacked. The large hoofed vertebrate had arrived in Lisbon enveloped, protected, and hidden by its robust skin, rumored to be that metallic or scaly armor with reptile residues or accretions that was to be immortalized by the German master. But it did not just have a skin. It also had a name known to all: rhinoceros. In fact, until it arrived in Lisbon that is all it was: name, verb, predicate. What people knew about it was part legend or invention, part certain and true. Strabo and Pliny had described it. Its forms, its habits, its rivalry with the elephant preceded it; they had determined both its fate (the duel in the Terreiro do Paço, its voyage to Rome) and its appearance in the popular imagination, once it had been seen, portrayed, and mechanically reproduced. Words had come before image, given the animal a shape, configured it. The reader will recall that the caption dominated the image in Dürer's woodcut.

Our second protagonist, on the contrary, lacked both words and predicates. It had no history. It had never been described. Little was known about its appearance; the only clues were its apparently contradictory bones with the paradoxical incompatibility of the teeth and the claws. Even less was known about its habits and practices. There was no narrative, legend, or fable that preceded it, nothing to help it take shape and come to life. It was a skeleton without flesh, skin, or discourse, an "unspeakable monstrosity" as Garriga had called it.[2] It was an animal without attributes, something that could not be named, merely a bundle of unconnected bones.

Otto Brunfels, one of the great botanists of the sixteenth century, called new botanical species that were not included among the six hundred described by Dioscorides "naked plants" (*herbae nudae*). That is precisely what the skeleton of the quadruped from the Río de la Plata was like: bare bones, washed clean of legends and stories. Like the plants that had not been described in the ancient world or the fauna of the New World discovered in the sixteenth century, the bones were unprecedented and impossible to classify—naked, mute, silent bones, petrified by time, lacking words, flesh, skin, and history. And without history, they had no name or identity either.

The creature needed a name and a place in the chart of natural history. But its strange anatomy and its size presented a challenge to nomenclature and taxonomy, the two classic instruments of natural history. A Madrid taxidermist was given the task of reconstructing and assembling it. There is a clear connection between taxidermy and taxonomy (both are derived from the Greek ταξις, meaning "ordering, arrangement"). The problem of its identity was connected with order and place. To put it in its place and give it a skin, to give it a name and a place in the animal kingdom, it was first necessary to connect, assemble, and mount that ensemble of disjointed bones.

So before it could be named, described, and placed in its appropriate zoological niche it had to be assembled and given form. First, it had to be given an image, in other words, *imagined*. In spite of its imperfections, the assembly by Juan Bautista Bru was an important step in that direction. It enabled the creature to assume a form and unleashed the process of inquiry that would finally lead to its identification and the solving of the enigma.

We have seen how the taxidermist cut, sawed, and filed as he chose. In his defense it must be said that this was the first time ever that anything like this had been done in a museum. These were the days before mammoths, mastodons, and dinosaurs stood before us. There, in the Royal Cabinet of Natural History in Madrid, the first reconstruction of an entire extinct animal was taking place. Once it had been put together, the drawings and engravings of the parts proved to be even more effective than the mounting itself, because they allowed the image to circulate and introduced the case to the scientific community at large.

Thanks to both activities—the sculptural work of mounting the skeleton and the pictorial one of engraving it—the vertebrate bones could be released from the soil that covered them, assume their shape on the pedestal of the Royal Cabinet, and circulate beyond Madrid. After having been buried for thousands of years, the animal was reborn. If the art of Dürer and the technique of engraving had formalized the image of the rhinoceros, multiplying it and reproducing it in space and time—immortalizing it in a certain sense—science and art secured a similar miracle with our second protagonist: its resurrection.

IT IS INTERESTING to consider some of the roles performed by bones in medicine and natural history and some cases of their visual representation in the early modern period. Andreas Vesalius (1514–1564), the great Renaissance anatomist, lent dignity to their study. Physicians had not taken much interest in bones before then. The study of bones was considered secondary, prosaic, mundane, and was left until the final year of university study. Vesalius, however, had the audacity to dedicate the first book of his famous treatise *De Humani Corporis Fabrica* to them in 1543. For him, bones represented the structure of the human edifice (Figure 20). The bones were the hardest, driest, coldest, and most earthly part of the organism. Their role in human movement and structure was like that of walls and pillars in architecture, an idea that he took from an influential treatise by Leon Battista Alberti. As for his system of labeling his drawings of bones with letters that recur in the body of the text, these were taken from a treatise on architecture by Sebastian Serlio.[3]

Before the bones could be seen, observed, and studied, the body had to be opened. Its interior could then be investigated, like the exploration of the globe. For the three centuries between 1500 and 1800, surgeons, physicians, naturalists, and travelers dissected their respective terrains: the body and the earth. Modern science provided the scalpels, chronometers, and sextants to explore their length and breadth; the objective was to establish the anatomy of the world and the cartography of the organs and entrails. This is not a conceit, a metaphor, or a historiographical analogy. Many anatomical atlases of the early modern era display this parallel between operation and exploration, such as the frontispiece of an edition of the *Theatrum anatomicum* of Giulio Casserio (1656) (Figure 21). It contains a cadaver whose abdomen has been opened up, an *écorché*, a skeleton, and a terrestrial globe showing America—all new worlds that had come into view thanks to the art of dissection and the voyages of discovery.

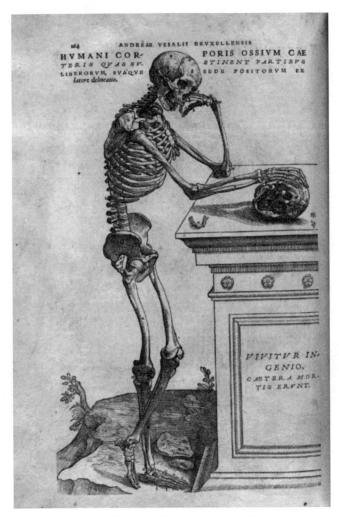

Figure 20. Andrea Vesalius, *De Humani Corporis Fabrica* (1543).

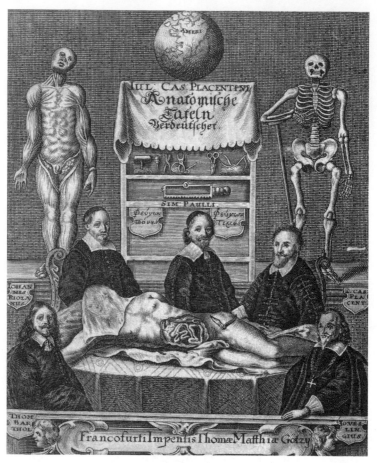

Figure 21. Giulio Casserio, *Theatrum anatomicum*, in *Anatomische Tafeln* (1656).

In fact, one of the principal motives of all modern science was precisely this: to reveal invisible things.[4] Showing the true nature of things, discovering hidden phenomena, visualizing bodies that had not been seen before, all took hold of the rhetorical and programmatic declarations of modern science from Galileo to the *Encyclopédie*.

Microscopy, the new technique of observation that was promoted from the second half of the seventeenth century, played a large part in this. Robert Hooke, who conducted experiments with the microscope in the early days of the Royal Society of London and wrote one of the first treatises in this field (*Micrographia*, 1665), was one of those extraordinary individuals who managed to broaden the outlook of his contemporaries and his successors.[5]

Hooke turned his microscope on all kinds of materials, revealing the wonders hidden within apparently trivial objects and creatures: the tip of a needle, taffeta fabric, the eye of a fly, a piece of wood, skin, bones. But we are less interested here in Hooke's contribution to osteology or his work on fossils than in his contribution in the field of the scientific image. In the preface to his *Micrographia*, Hooke considered the role of the instruments of observation that remedied "the infirmities of the Senses." Thanks to the telescope, nothing was too distant to be brought into view; thanks to the microscope, nothing was too small to escape examination. In his view, what was needed to reform philosophy was simply to observe those images:

> In this kind I here present to the World my imperfect Indeavours; which though they shall prove no other way considerable, yet, I hope, they may be in some measure useful to the main Design of a reformation in Philosophy, if it be only by shewing, that there is not so much requir'd towards it, any strength of Imagination, or exactness of Method, or depth of Contemplation (though the addition of these, where they can be had, must needs produce a much more perfect composure) as a sincere

Hand, and a faithful Eye, to examine, and to record, the
things themselves as they appear.[6]

As a good representative of modernity, Hooke preached a
turn from ratiocination and imagination to observation and mi-
mesis, from hearing to the hand and the eye, in other words,
from rhetoric and the word to pure facts and their images. The
instruments of observation brought objects closer or magnified
them. They produced creditworthy images inasmuch as they
generated them artificially and mechanically, which recalls the
effect of printing on the word and of engraving on the image.
Modern science is that devious process that begins with a eu-
logy of sensory experience and the *ad vivum* image before
slowly but invariably turning toward a defense of anesthetized
knowledge that is mediated by machinery: technologically
objectified.

Claude Perrault spoke in similar terms about images in his
Mémoires pour servir à l'histoire naturelle des animaux (1673–
1676). Their purpose was to represent things exactly as they had
been observed, "as in a mirror, without introducing anything
of its own and representing only that which is presented to it."[7]
In fact, from the days of Leonardo and Dürer down to our
fossil, the scientific image had been gaining a reputation as a
universal language that eliminated the confusion and verbal
categories of language proper. As the anatomist William Hunter
(1718–1783) was to remark in a lecture to the Royal Academy of
Arts in 1770, "Representation in the imitative Arts is a Substi-
tute for reality."[8]

Thanks to the perfection of the techniques of metal engraving (those concerned with the uses of the burin and the
etching), reproductions of bones in anatomical atlases achieved

an unexpected quality, resembling photography. In fact, William Cheselden, author of one of the best osteological atlases of the modern era (*Osteographia or the Anatomy of Bones*, 1733), used the camera obscura to achieve extraordinary results. He also had the bright idea of including some animal skeletons with the human ones, setting up a visual comparison, and implicitly suggesting the family resemblances between vertebrates to the observer (Figure 22).

Other highlights of osteological illustration were the treatises by Govert Bidloo (1649–1713) and Bernhard Siegfried Albinus (1697–1770), two important anatomists whose engravings clearly follow the *vanitas* tradition and other pictorial formulas regularly used to represent human remains in the late Renaissance and Baroque periods. Bidloo studied under another singular physician, Frederick Ruysch (1638–1731), whose dioramas and sculptural compositions with skeletons and human remains were macabre enough to be acquired by Peter the Great. As for Albinus, he and his engraver, Jan Wandelaar, perfected the grid system devised by Dürer to project the proportions of bodies and osteological forms. Among the splendid illustrations in his *Tabulae sceleti et musculorum corporis humani* (1747) are two that must be mentioned here: the front and rear views of a human skeleton in a scene dominated by the presence of a live rhinoceros, also represented in frontal and dorsal views. This successor to Ganda was the famous rhinoceros that was taken to the Netherlands and to cities in central Europe and Italy between 1741 and 1758. It is used here, as in other illustrations of the same atlas, to create a stagelike setting and for its tonality: the scale of grays of the thick hide adds depth and contrasts strongly with the skeleton, which is superimposed in its delicate brightness against the bulk of our hoofed pachyderm (Figure 23).

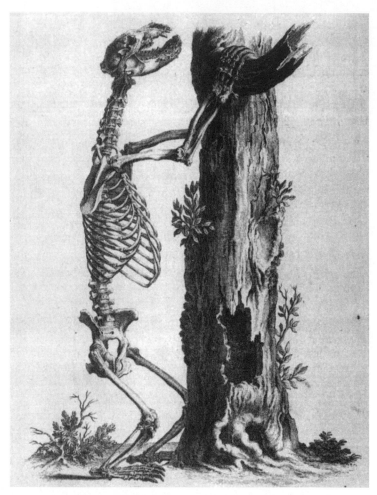

Figure 22. William Cheselden, *Osteographia or the Anatomy of Bones* (1733).

Figure 23. Bernhard Siegfried Albinus, *Tabulae sceleti et musculorum corporis humani* (1747).

The metaphysical weight of the skeleton, especially the human skull, had a considerable influence on reproductions and osteological atlases. Like Hooke, Crisóstomo Martínez (1638–1694), another classic practitioner of Baroque anatomy, serves to illustrate the suggestive intersection between osteology and microscopy. The allegorical content of his work is indebted to the pictorial tradition that associated bones with relics, transcendence, and the futility of life.[9] Paradoxically, the most mundane and earthly part of the organism turned out to be the most permanent, which suggested eternity in a more direct way.

Hooke's "sincere Hand" and "faithful Eye" sometimes went their separate ways; while the eye remained the prerogative of the physician, the hand could (or should) belong to someone with very specific skills, an artist trained to record the forms and profiles of nature. The formula had been successfully employed by Vesalius in *De Fabrica*, in which he made use of various artists; among them was Jan van Calcar, a pupil of Titian. The division of labor was the logical result of specialization, the development of individual disciplines, and the perfection of the arts, although it also expressed the traditional division between mental and manual labor. This did not prevent many from practicing both. Leonardo and Galileo, who were both artists and scientists, were exceptional less for combining these skills than for the tremendous talent that they displayed in both fields. For anatomy, there have always been physicians and surgeons who were also excellent draftsmen. Physicians, moreover, routinely collaborated with master draftsmen, who were practically indispensable in all the osteological atlases.

The alliance between art and science was fertile and became routinized to the point of becoming a specialization in its own right. Artistic anatomy was a combination of practices and requirements to be found in the curriculums of the art academies of the eighteenth century. Dürer, author of a treatise on the proportions of the human body, and Leonardo had been the great precursors of this discipline.

Treatises and manuals of artistic or pictorial anatomy proliferated in the Enlightenment. We have, for example, Charles Monnet's *Etudes d'anatomie à l'usage des peintres* (1770–1775) with engravings by Giles Demarteau, and *Elémens d'anatomie, à l'usage des peintres, des sculpteurs, et des amateurs* (1788) by Jean-Joseph Suë the Younger, professor of artistic anatomy in the Academy of Painting and Sculpture of Paris. His father had taught in the College of Surgeons and in 1759 had translated another of the best osteological treatises of the eighteenth century, Alexander Monro's *Anatomy of the Humane Bones* (1726). Suë junior continued to use this work for his lessons in artistic anatomy. It was very common for the plates and engravings of bones to pass from one edition to another. For example, those of Albinus were used in the treatise on artistic anatomy by Ploos van Amstel (1783) and later in that by Johann Heinrich Lavater (1790), which was translated into Spanish as *Elementos anatómicos de osteología y miología para el uso de los pintores y escultores* (1807). In the prologue of the latter, the translator, Atanasio Echevarría y Godoy of the Royal Academy of San Carlos in Mexico, recalled the necessity of treatises on anatomy for young art students. Painting and sculpture were "arts that imitate nature"; their perfection lay in producing "copies that were very close to the originals."

It was only natural that animal bones should receive much less attention than human ones. Throughout the early modern era, they occupied only a marginal chapter in the atlases of zoological and natural history. Exceptionally, they had featured in an atlas of human osteology, as in the case of Cheselden mentioned above, and in the work of artists who specialized in representing animals, such as George Stubbs, the painter of horses who produced one of the first representations of equine bone structure in his *Anatomy of the Horse* (1766). However, the systematic study of animal skeletons had to wait for the coming of comparative anatomy, an emerging discipline that made use of the study of the fossil remains of large vertebrates, the field that was born—so to speak—after the discovery of the skeleton from the River Luján. By one of those strange shortcuts in the history of science, zoological osteology and vertebrate paleontology (two modern terms that are used here to refer to fields without a name or a space at the end of the eighteenth century) both found themselves in the eye of the storm in one of the greatest renewals in the life sciences since the days of Lucretius.

Thus some of the best illustrations of zoological osteology (there were not many) originated in a field that was small and vaguely defined. The interest in animals' fossilized bones was cultivated by a minority of scientists in the last third of the eighteenth century—naturalists, students of fossils, and anatomists interested in certain pieces that were scientifically attractive for two reasons: they were strange (singular, unique), and they were unprecedented (they had not been represented before). Among them are an exemplar from Maastricht (a full-fledged chimera with features of a whale and a crocodile) and the remains of large quadrupeds found in Siberia and Ohio, which were assumed to come from elephants, rhinoceroses, or related species. The best way to communicate these finds, in fact the only way to publicize them amid the speculations and hypotheses that surrounded them, was to represent them "with a sincere

Hand and a faithful Eye." These bones, like those from the River Luján, had no tradition, no authority to resort to. They belonged to unspeakable monsters. They were naked bones, inscrutable testimonies. The first thing to be done—not the only one, but the most important—was to reproduce them exactly and to circulate them. The latter was possible thanks to the long-distance exchange of drawings, a practice associated with correspondence between men of science, and the engraved plate, the instrument for fixing and circulating the same facts among the global scientific community. Bones, like many other natural products, machines, procedures, or instructions, became portable and replicable—universal—thanks to the intensification of the exchange of correspondence, but above all thanks to advances in the art of engraving and the rise of periodical publications and the publishing industry. If the early modern world began with the invention of movable type and the first circumnavigation of the earth, the Enlightenment was bringing both tasks to their completion with the exponential expansion of the Republic of Letters and the encirclement of the globe with dense networks of trade, travel, and mapping. The *Encyclopédie* and the voyages of Captain James Cook stand out for the forceful way in which they embody both impulses: education and circularity.

ALL THIS SHOULD help us to better see and understand the images of our skeleton. On the one hand, in the absence of an autonomous pictorial tradition, animal bones would inherit certain conventions of the characteristic representations of human bones. On the other hand, in their role as curiosities and regular items in cabinets and museums they also received a formal

treatment like that accorded other fossils and natural products, engraved and standardized in those Enlightenment catalogues that were dominated by symmetry and the search for
regularities among the forms of nature. The representation of
animal bones consequently combined elements of human osteology, the architectural treatise, the anatomical atlas, and the
vanitas, but also something of the cabinet of curiosities, in other
words, the still life exposed to that geometrical treatment characteristic of the Enlightenment that converted everything it
touched into an object to be placed within a system.

As mentioned in Chapter 4, the drawings and engravings of
the beast from the River Luján were the work of Juan Bautista
Bru, taxidermist and painter of the Royal Cabinet of Natural
History, and Manuel Navarro, an engraver associated with the
San Fernando Academy of Art (founded in 1752), which was
housed under the same roof as the cabinet where the mysterious animal was kept. Three new chairs were created in this
academy at the end of the 1760s on the initiative of Felipe de
Castro and Raphael Mengs. These were for geometry, perspective, and anatomy. The physician Agustín Navarro was appointed professor of surgery and anatomy in the academy in
1768.[10] Bru had been taught there before becoming the cabinet's official taxidermist. He had learned to draw the bones,
skull, and musculature of the human being in the exercises that
students carried out with corpses from the nearby general hospital. Despite specializing in anatomical drawing, Bru failed to
be admitted as a member of the academy.[11] In any case, studies
of that kind were slow to become consolidated and toward
the end of the century were in a perilous state. The academy,
the Royal Cabinet, and many other educational and scientific
institutions in Spain had inadequate preparation to confront

the stormy years of the Revolution and the war. They were too young and lacked what every institution requires: an academic tradition, a couple of generations of a qualified staff, good practices, and financial support, that combination of requisites that is so fertile and so rare in the history of Iberian science.

All the same, the academy enabled print techniques, particularly etching, burin engraving, and their combination in the technique called "*talla dulce*" (in which the burin is used on a soft-ground etching), to take root in Spain by conferring on them the same dignity as the other plastic arts.[12] The year 1789, in which the skeleton was being assembled (while the monarchy and the entire ancien régime political body were being disassembled in neighboring France), witnessed the setting up of the Calcografía Nacional as an annex to the Royal Printing House to work for the various secretaries. This new establishment printed plates to illustrate the *Guide for Foreigners*, the *Portraits of Famous Spaniards*, and *The Views of Seaports*, works whose printing and circulation were intended to standardize the images of the places, people, and monarchs who constituted the iconography of the nation, although its principal task, more routine, was to print the *vales reales*, a sort of paper currency created by Charles III to deal with royal debt in a time of political crisis.[13] The art of engraving was thus at the service of the emission of the public debt of the Bank of San Carlos (the predecessor of the National Bank, another recent creation). The art of the engraving permitted the multiplication and distribution of identical images in three different spheres—the credit system, the iconography of the nation, and science. It thereby generated what a community needs to guarantee its cohesion, to ensure that everybody, no matter how isolated, uses the same frame of reference. If trust is the basis of the market, the credit system and political representation, social cohesion and the exchange of information, then engraving was the tool that could generate it on a large scale. The uniformity of the gazes and the collective imagination proceeded via the standardization of *imagined* objects.

When Bru produced his drawings of the skeleton, he already had other "scientific" works behind him, principally the illustrations of his "Collection of plates representing the animals and monsters of the Royal Cabinet of Natural History of Madrid" (1784–1786) and some of the iconographical works for the two projects led by Antonio Sáñez Reguart, war commissioner of the navy: the unfinished "Collection of fish and other maritime products of Spain" and the "Historical dictionary of the national techniques of fishing" (1791–1795).[14] The "Collection of plates . . ." was a zoological work resembling in some ways the classic works of Gessner or Jonston, but containing more up-to-date information from Linnaeus and especially from the Comte de Buffon. The illustrations were hand-colored plates. Bru claimed to have done the original drawings from "life," that is, from the dried exemplars in the cabinet. His images have a certain attraction because they have the naïve, primitive air of the iconography of early modern natural histories. They bear no relation to the resolute precision and realism of the severe profiles and extraordinary shading of the preparatory drawings and the engravings of the skeleton assembled in the gallery of petrifications of the Royal Cabinet. When compared with a critical eye, it is hard to believe that both sets of illustrations were made by the same hand.[15]

As for his role in the works coordinated by Sáñez, Bru was in charge of the team that engraved and colored the original drawings for the superb ichthyological collection. It remained un-published owing to a series of complications that are not rele-vant here, but which have been the subject of a polemic between specialists and even served as the plot for a recent novel.[16] Soon afterwards, Bru supervised the drawings and preparatory work on the dictionary of fishing techniques, a unique work of its kind on which various artists and engravers collaborated.[17]

One of these was Manuel Navarro, the engraver of the plates of the skeleton. Born in Zaragoza, Navarro came from the Ara-gonese school of engraving, although by this time he already belonged to the Madrid circle of Manuel Salvador Carmona, the most important figure in Spanish engraving of his day.[18] Director of engraving in the Academy of Art since 1777 and engraver in the king's chamber from 1783, Carmona had intro-duced the sophisticated technique of *talla dulce* (as noted, a mix of etching and engraving), which he had learned in the studio of Nicolas Dupuis in Paris. In spite of the low artistic es-teem in which engraving was held at the time, partly corrected by its institutionalization in the academy, it was complex work whose techniques took more than a decade to learn and de-manded a huge physical effort, to the extreme that it left Sal-vador Carmona blind, like many a microscopist, surgeon, and conchologist of the day: they all overworked the organ on which the experimentalists and craftsmen depended.[19]

Navarro engraved numerous prints of the Madonna and saints as well as participating in collective works of a certain stamp. He was responsible for many of the plates in the "Description of the imperial canals of Aragon and Real de Tauste" (1796), a classic of the Enlightenment in Aragon. He also worked on the Spanish edition of one of the last major works on emblems, the *Iconología* of Gravelot y Cochin (1801), whose fame rests on one of the most famous engravings ever made, the frontispiece of the *Encyclopédie* in which Truth is revealed by Reason and Philosophy in all its dazzling beauty (Figure 24). Truth as the conquest of *logos*, or as the inexorable epiphany of nature, is an image that radiates so much light that it is impossible to see her for what she is: an image illuminated and multiplied by the engraving, an artifice that obtained its aura outside time—its sacrality—thanks to the technique of engraving and mechanical publishing.

Navarro was therefore a highly qualified engraver, as his plates of the skeleton clearly show. Their excellence is also due to the splendid typography used by the press of the widow of Joaquín Ibarra. The question of how the cooperation between Bru and Navarro proceeded and who was responsible for these engravings is again very delicate. Bru has had his supporters and his detractors. He has been hailed as the unjustly overshadowed precursor of vertebrate paleontology and as the author of some of the best scientific illustrations of eighteenth-century Spain. The first claim is completely unfounded; the second is accurate enough if the illustrations really are his, for he was already accused of plagiarism, parasitism, and forgery during his own lifetime—someone who made a habit of living off the work and merit of others.

This, however, is not relevant to our argument. Whoever their author, these five plates are a technical wonder and a work of beauty. They were created with the *talla dulce* technique, the usual technique used among the students in the academy and the artists of the youthful Calcografía Nacional. This technique was slowly but surely replacing the woodcut in the course of the

Figure 24. *Encyclopédie*'s frontispiece, drawing by Charles-Nicolas Cochin in 1764, and engraving by Benoît Louis Prévost in 1772.

early modern period. Navarro must have learned it (or at least mastered it) in the studio of Salvador Carmona, who had introduced it in Spain.

The pressure of the hand on the burin dictated how deeply it cut into the copper. With great dexterity, the index finger guided the chiseled tip of the instrument, while the hollow of the palm was opened or closed to graduate the intensity and to regulate the depth or subtlety of the groove. By means of a varied repertoire of wavy, parallel, or hatched lines at varying distances from one another, Navarro obtained the shading and graduation of the volumes, the extremely faithful representation of the bone cavities. The precision of these images is tremendous. Of course, burin engraving required a highly skilled draftsman. All the engravers of this level were familiar with the work of the great masters and with the principles of perspective and geometry. They knew about architecture and anatomy.[20]

The best known of the five plates is the one of the mounted skeleton. It is a priceless image as the testimony of the first reconstruction of an extinct animal ever (see Figure 19). It reproduces the mistakes already made during the assembly of the bones, in line with those committed during its first reconstruction in Buenos Aires. The animal's posture on all fours shows what Bru and his assistants had in mind: a pachyderm, a big herbivore, or perhaps a colossal carnivore. Navarro engraved his name in the bottom right corner; the name of Bru is engraved on the left, surmounted by a scale in French feet and one in Spanish rods, representing the metrological diversity that the Revolution was going to end. The preparatory drawing for this plate is at least as interesting: with the characteristic grid to reproduce the proportions in scale, the image reflects one of the

Figure 25. Juan Bautista Bru, preparatory drawing for plate number 1 in José Garriga, *Descripción del esqueleto de un cuadrúpedo muy corpulento y raro, que se conserva en el Real gabinete de Historia Natural de Madrid* (1796) (see Figure 19).

audacious touches that Bru carried out during the reconstruction—the addition of the mule's tail (Figure 25).

The other four plates reproduce the separate bones grouped by anatomical criteria. They are sections like those in osteological atlases.[21] The image of a skull, formidable in many respects, occupies the center of one of them (Figure 26). Owing to its volume and form, the way in which the bony depressions, the apophyses, and the orbital cavities are reproduced, this image of a mute and portentous skull, as silent as it is inexplicable, is a zoological *vanitas* that captures a unique moment on the eve of the history of the earth and of life. It is a startling document bearing on certain truly revolutionary ideas and facts.

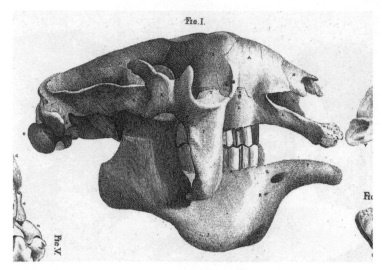

Figure 26. Juan Bautista Bru and Manuel Navarro, detail of plate number 2, published in José Garriga, *Descripción del esqueleto de un cuadrúpedo* . . . (1796).

PERHAPS BRU WAS the author of these drawings, and therefore the engravings should be largely attributed to him too. However that may be, he was not the Dürer of this history. He was not the one who resuscitated and baptized the skeleton. Bru was an intermediary between Manuel de Torres, who had unearthed the bones in the Río de la Plata, and a young naturalist who had just arrived in Paris. His name was Georges Cuvier (1769–1832).

As in the case of Dürer and his rhinoceros, Cuvier brought the skeleton to life without having seen it, thanks to copies of the original drawings and engravings. The role of the taxidermist Bru in this history is closer to that of the anonymous author of the drawing of Ganda. The role of the Moravian merchant,

Valentim Fernandes, who sent that drawing of the rhinoceros from Lisbon to Dürer in Germany, is replicated by the French diplomat Philippe Rose Roume, who mediated between Bru and Cuvier. Among other things, this episode demonstrates the multiplicity of actors who intervene in the production of knowledge. Science is a collective, social activity. A scientific fact is not established by a single eye or a single hand. More or less faithful, sincere, delegated, or imposed—the number makes for the variety—the eyes and the hands that produce facts (not to mention knowledge) are much more numerous than one might imagine.

Roume was in Madrid at the end of 1795 to negotiate the terms of the cession of Santo Domingo to France. He was familiar with Caribbean affairs, being himself a native of the island of Granada. He was also interested in the natural sciences, so when he visited the Royal Cabinet he could not help being captivated by the majestic skeleton. He obtained copies of the engravings of Bru and Navarro and sent them to the recently created French Institute in Paris, where the young Cuvier resolved the mystery of the beast's identity and gave it a name in a flash of—what? Lucidity, mastery, genius, method, scientific rigor, imagination?

A little of each. And the good fortune of being in the right place at the right time. Cuvier had just turned twenty-six. He was born in Montbéliard in the Duchy of Württemberg and was therefore completely bilingual (he was later to master many more languages besides French and German). The scion of an established Protestant family, he had studied for an administrative career in the Carolinian Academy in Stuttgart, although from his boyhood on he had been passionately interested in natural history. During the most turbulent years of the French

Revolution he lived on the coast of Normandy, where he studied marine fossils whenever the fancy took him while earning a living as tutor to a noble family. After the Terror was over he settled in Paris in 1795, where he was appointed to a position in the Museum of Natural History and soon afterwards in the French Institute, the two institutions with which the Republic wanted to lead science in the new era.[22]

The Royal Garden and the Royal Cabinet, two jewels in the crown whose bearer's head had recently rolled, were merged with the Museum of Natural History.[23] The magnificent collections of animals, products, and naturalia (live and stuffed animals, remains, skeletons, and the largest collection of animal bones in the world) would allow Cuvier to work in a way that would have been impossible elsewhere. At the beginning of 1796 he had just obtained the position of assistant to a professor of animal anatomy, the venerable Antoine Mertrud. The museum was dominated at the time by such figures as Louis Jean Marie Daubenton (1716–1800), its director, a physician, anatomist, expert on vertebrates, and close collaborator with the great Buffon, who had died before the taking of the Bastille. When the drawings of our skeleton arrived in Paris in those early months of 1796, Cuvier was the latest to arrive in the museum and was completely unknown there. He was not even the youngest or the most promising; that was the brilliant Étienne Geoffroy Saint-Hilaire (1772–1844), professor of zoology, with whom he was soon to collaborate and later to disagree.

The National Institute of Science and Art was a genuine product of the Revolution. The Directory had founded it only a few months earlier, in October 1795, shortly before Roume's visit to Madrid. The institute wanted to combine the functions of the old royal academies (of science, inscriptions and litera-

ture, painting and sculpture, and the rest) that had all been sup-
pressed by the French National Convention in 1793 as—in the
words of Abbé Grégoire—"gangrenes of an incurable aristoc-
racy." It was created to repair the damage inflicted on French
culture and science in those heady years, to relaunch and cen-
tralize them, to regulate them, and, finally, to put them at the
service of the state. In this context, its activities were divided
into three areas and classes: the third for literature and the arts,
the second for moral and political science, and the first for the
natural sciences of "physics and mathematics." This meant that
natural history, botany, mineralogy, anatomy, and even chemistry were now to some extent subordinate to the subjects that
had been protagonists in the Scientific Revolution, Newtonian
science definitively consecrated by Laplace. They were the sciences that had been practiced in the Royal Academy of Sciences—one of the oldest of its kind in Europe and the most
outstanding throughout the eighteenth century—which had
recently been abolished and was now subsumed under the new
institute. Cuvier was assigned to the first Class of Physical and
Mathematical Science, a title that throws light on the ambition
of converting zoological osteology into a kind of geometry or
mathematical discipline.

The copies of the engravings of Bru and Navarro sent from
Madrid by Roume reached Abbé Grégoire, a protagonist in the
Revolution and a dominant figure in the new scientific policy
and in the institute itself. He passed them on to Cuvier and
asked him to write a report on the case. At the sight of the
images of Bru and Navarro, the new arrival used the materials
available in the museum where he lived and worked to write a
report that gave rise to a formal intervention and to his first article on fossil osteology and comparative anatomy, the first in a

long series of researches that would eventually set him up as one of the scientific luminaries of the first half of the nineteenth century, on a par with Humboldt or Darwin. It is worth reproducing its title in full: "Notice sur le squelette d'une très grande espèce de quadrupède inconnue jusqu'à present trouvé au Paraguay, et depose au Cabinet d'Histoire Naturelle de Madrid." It was published in the *Magasin Encyclopédique* and immediately translated into English in the *Monthly Magazine* in that same year, 1796.[24]

Found in Paraguay? Cuvier may have known a lot about bones, but he was ill-informed on South American geography. The slip reveals the characteristic disdain with which the peripheries are regarded by those who know that they are at the center (to confuse Uruguay, on the other side of the River Plate, with Paraguay, is a glaring error), but more important it shows us that Cuvier did not know much about the context in which the cadaver had been found. It simply did not interest him. Nor was he impressed or confused by its size, among other reasons because one of the comparative advantages of the drawings, in relation to the skeleton, was that they were scale reproductions. On paper the dimensions of the animal become something relative; they are minimized or even disappear from sight altogether, making it easier to appreciate the morphological parallels without taking the dimensions into consideration.

Cuvier concentrated on the forms and resolved the paradox of the herbivore's teeth and carnivore's claws by using his knowledge of other vertebrates and applying it to the case at hand by analogy and extrapolation. This was the first occasion on which he put into practice his two major anatomical principles: the correlation of the parts of an organism, and the subordination of characters. All the organs of a living creature and all the parts of these organs were related to one another and depended on the functions that they performed: in other words, functional structuralism, or morphology. For example, a carnivore needs claws to catch its prey, incisors to be able to chew it, and a digestive system capable of digesting it. Osteology became a game of logic, a practice of forensic science: give me a few bones and I will reconstruct an individual, give it a skin, shape, habits, life. Nature never acted in vain—as the classics had surmised—and so to determine the forms and habits of a living creature was a matter of keen observation and the application of logic.

In the present case, the first thing Cuvier had to do was to validate the plates as trustworthy images and to assume that they realistically represented and could act as substitutes for the actual bones. He had to be sure that examining the plates was equivalent to having the bones in front of him (Figure 27). Although the illustrations of individual bones must have been more useful to him than that of the assembled animal (what he called "a bad copy of the figure of the whole skeleton"),[25] the latter fit with his visual rhetoric, a fundamental part of his scientific argument and expository program, so he decided to reproduce the image of the reconstructed skeleton, which through the process of engraving is naturally reversed (Figure 28).

He drew two surprising conclusions: the bones did not match those of any known living creature and only resembled those of certain exotic species much smaller in size, the edentates, a family that includes the pangolin, sloth, armadillo, and anteater. In the whole of the animal kingdom these were the only ones to combine the peculiar enamel-free molars that keep on growing with the nonungulate extremities and the powerful claws that had confused other scientists so much. The lack of an

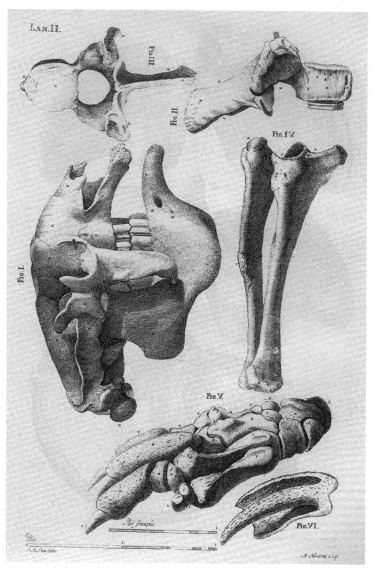

Figure 27. Juan Bautista Bru and Manuel Navarro, plate number 2, published in José Garriga, *Descripción del esqueleto de un cuadrúpedo . . .* (1796).

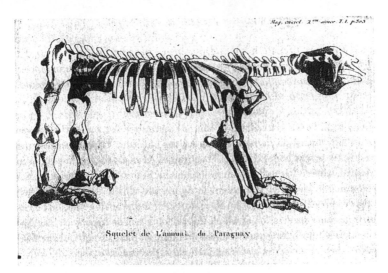

Figure 28. Georges Cuvier, *Megatherium americanum*, plate published in "Notice sur le squelette . . . ," *Magasin Encyclopédique*, 2e année, 1, 1796.

ischium and pubis for the pelvis (they had been lost during the exhumation) created difficulties, but the presence of clavicles, the width of the surfaces of the lower mandible, and the large apophysis at the basis of the zygomatic arch were characteristics of the edentate family. The length of the front legs was typical of the sloth; the robustness of the hind legs was a characteristic of the pangolin; and its teeth resembled those of the armadillo. That impossible monster thus followed the anatomical blueprint of several tropical species that were little known in Europe at the time and in any case smaller, much smaller, than the massive exemplar in the Royal Cabinet.

But how was one to imagine a sloth more than four meters long and two meters high? In Cuvier's opinion, the skeleton was that of an extinct animal completely different from any that "we can see today on the earth." It belonged to an old world *(ancien monde)*. But *how* old was that world, and *how* did it differ from the present one?

For the time being, Cuvier stuck to the evidence and to the questions immediately confronting him. He classified the specimen between the sloth (from the shape of its skull) and the armadillo (from its teeth) (Figure 29), and he ventured to call it *Megatherium Americanum*, "the great American beast." Soon afterwards he changed the name to *Megatherium fossile*. The case was closed. By applying the binomial nomenclature to a fossil, Cuvier was pointing to the existence of a remote past and a fauna that had disappeared from the face of the earth. Linnaeus, the modern Adam who had given new names to the living world when he formulated his binary nomenclature—a method devised practically as a mnemonic formula, a game to aid the memory—had said that he felt as though he was attaching the clapper to a bell. That is what happens when the right word is found to designate or name an object or a thing. Something resonates inside it. Cuvier must have had the same sensation. He had given a name to the strange cadaver and assigned it a niche in the animal kingdom, the first step toward recovering its skin: *taxonomy, taxidermy.* Thanks, first, to the image and then to the word, those bones were beginning to come to life and to recover their place and appearance. The giant sloth was rising energetically from its long, long winter sleep.

WHEN YOU ARE trying to do a jigsaw puzzle, you try out the pieces in different places. Something similar happens with social or natural facts. People measure and observe how they fit with one another and what the relations are between them.

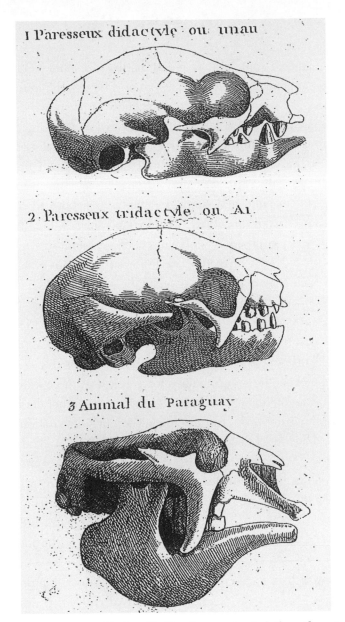

1 Paresseux didactyle ou unau

2 Paresseux tridactyle ou Aï

3 Animal du Paraguay

Figure 29. Georges Cuvier, Skulls of a two-toed sloth, a three-toed sloth, and Megatherium, illustration published in "Notice sur le squelette . . . ," *Magasin Encyclopédique*, 2e année, 1, 1796.

Vertebrate paleontology is a fascinating discipline. Perhaps more than jigsaw puzzles, which can only be solved in one way, it reminds us of those other construction games in which the pieces can be fitted together in different ways. It also reminds us of history. In all these practices, we work with material that is more flexible and reversible than it appears at first sight.

The Megatherium was slow to take on a stable form; or rather, it took on several. For a long period, but above all during the early years of this, its second biography or new life, its versatility was proverbial. After having been dead for thousands of years, its sudden resurrection was followed by various metamorphoses, changes of aspect, and adaptations to distinct scientific and political programs. We are looking at a subject that, like a good modern, resisted being stereotyped. It was ready to make up for lost time and to live not one life but several.

The first was an abortive attempt. It all began with the intervention of another intermediary, for in this story there were two "Moravian merchants." Long before Roume, another diplomat had anticipated him and circulated certain information and some drawings of the mysterious skeleton in the Royal Cabinet. This was William Carmichael, a North American agent who had collaborated with Benjamin Franklin on his Paris mission during the War of Independence and who was the official representative of the interests of the nascent United States in Spain between 1782 and 1795. He was also a devotee of the natural sciences.

Carmichael had seen the skeleton in the Royal Cabinet as early as January 1789, only a few months after the arrival of the bones and while the reconstruction was still under way. It was then that he wrote to Thomas Jefferson, giving details of what was happening in Madrid (most concerning the policy of

Charles IV toward the American nation). In that same correspondence he included "a description of the skeleton of an
animal recently discovered in Spanish America,"[26] documentation that incorporated a diagram (a drawing) and some notes,
certainly preliminary and incomplete, but sufficient to arouse the
interest and imagination of a person like Jefferson. These drawings and notes were produced by a local informant whose name
Carmichael did not mention: Bru? The surgeon from the
Hospital del Buen Suceso? Manuel Navarro?[27]

Thomas Jefferson (1746–1826) was living at the time in Paris.
He was the leading representative of the United States in France
and must have received the information and drawings in his
residence in the Champs Elysées. Carmichael sent him these
materials because of Jefferson's reputation among the salons
and academic circles of Enlightenment Paris, the great metropolis where courtly culture reached its apogee shortly before
it was set alight on all sides. The future president of the United
States was a man of many talents: university reformer, frus-
trated architect, and admirer of Francis Bacon, Isaac Newton,
and John Locke, who played a by no means unimportant role in
various disciplines of the natural sciences.[28] He knew and wrote
about astronomy, botany, ethnography, and agriculture. He
was a member of the American Philosophical Society from 1780
and was its president from 1790 to 1815. His *Notes on the State
of Virginia* (1787) was one of the most influential American
natural histories of the period. His interest in paleontology also
occupied him, albeit intermittently, for many years. He was familiar with the Paris collections and the Royal Cabinet and was
in contact with the most expert naturalists. He had been a sparring partner of Buffon, his opponent in the so-called Dispute
of the New World (the polemic on the inferiority of American

nature), and was at odds with his successor, Daubenton, the man who was to become director of the museum when the young Cuvier arrived in Paris in 1795. Throughout his life, Jefferson did much to promote and popularize science in his country. For example, he supported the Museum of Natural History of Philadelphia, one of the first enterprises situated midway between spectacle, science, and art, which was the brainchild of a visionary, his friend Charles Wilson Peale, who later gave dinners inside the skeleton of a mastodon while a band played the national anthem.[29] During the presidency of Jefferson, a visitor to the east wing of the White House remarked on its walls covered with maps, globes, and books; the window-sills were laden with roses, geraniums, and other plants that Jefferson cultivated personally. The skeleton of a mocking bird (*Mimus polyglottos*), a typical bird of North America resembling a blackbird or nightingale, hung above them.[30]

For Jefferson, a lot depended on the bones of American fauna. Like other naturalists, especially those with an interest in the skeletal remains of the New World, Jefferson was fascinated by the discoveries in Big Bone Lick, marshland south of the Ohio River in Kentucky.[31] The first finds in this deposit went back to 1739, when some French soldiers had come across the bones of a strange specimen. They seemed to be those of an elephant, but soon began to be compared to those unearthed in the succeeding decades in various parts of Siberia, the mammoths. Excavations multiplied in both regions in the two hemispheres in the 1760s, accompanied by interest and debate: How could there have been elephants or similar animals in such northern latitudes? Were they polar relatives of the tropical pachyderms?

On the animal from Ohio "known" as *American incognitum,*
that is, "unknown American," depended not only one of the
most interesting paleontological debates of the period, but also
the reputation of a continent whose nature had been looked
down upon by half of educated Europe.[32] Following William
Robertson and the Abbé Raynal, with a more solid basis than
Voltaire and certainly with more substance than Cornelius de
Pauw, the Comte de Buffon had more than anyone else laid the
scientific foundations for a systematic thesis on the inferiority
of American nature: one resting on claims about the cold and
damp environment, the continent's immaturity, the decadence
of its domestic animals, and the absence of large forest dwellers.
The New World was precisely that: of recent origin, formless,
swampy, and unhealthy. Following an implacable Aristotelian
logic, Buffon considered bigger better than smaller, stability
better than mutability. Seen this way, the Ohio animal, considerably larger than the elephants of Africa or Asia, required another explanation. Buffon's ideas on geology came in useful
here. In *Les époques de la Nature* (1778) he had expounded the
theory of the gradual cooling of the planet, which he deduced
from changes in animals and the climate. The *American incognitum* might have been larger than other analogous Old World
species—there was no denying it—but that was because it was
an extinct animal (in fact, the only one that Buffon recognized
as extinct), a species that had lived in the remote past (in the
fifth epoch of his system), when North America had a much
warmer climate.

These theories were contested by various members of the
local élites who had been born in America: Creoles in the
Spanish viceroyalties and inhabitants of the British colonies

who were now citizens of the new republic. The slur affected them directly, of course, because the intellectual and emotional capacities of the Americans were at stake. Natural questions are never closer to social ones than when they bear on the human skin (remember Valéry's dictum "Ce qu'il y a de plus profond en l'homme, c'est la peau"). It is the surface where our emotional and intellectual capacities and our moral dignity emerge, or onto which they are projected. One of the main opponents of these theories was Jefferson, who turned the tables to make America exactly the opposite: a world blessed with the best natural products and landscapes. The land of the free and of the future was incomparably fertile, powerful, and robust. The animal from Ohio could not have been better suited to bolster his defense of American nature, since his providentialist and deist vision had no room for transformism or for the extinction of species.

Jefferson was on the verge of using the Megatherium for the same ends. As we have seen, he received reports and drawings of the Madrid animal from Carmichael at the beginning of 1789 (though without identification or a name and only half assembled). But then the Revolution broke out and he had to leave French soil to return to America before the year was out. His mind was on other matters for the next few years, but the specter of the *American incognitum* returned to haunt him in 1797, when the remains of another strange cadaver were discovered, this time in the vicinity of Greenbrier, Virginia. These were the bones of various limbs, including some magnificent paws armed with sizable claws. Jefferson rapidly showed an interest in the case but, contrary to his expectations, the animal found in Virginia was not like the gigantic

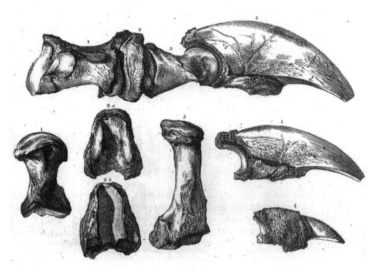

Figure 30. Caspar Wistar, fossils of claws of Megalonyx, "A Description of the Bones Deposited, by the President, in the Museum of the Society, and Represented in the Annexed Plates," *Transactions of the American Philosophical Society*, vol. 4, 1799.

elephant of Ohio, a mammoth (in fact, the beast of Ohio was later identified as a mastodon). The Virginia animal was not an ungulate. Instead of hooves, it had fearsome claws. Jefferson even acquired some parts of the skeleton (kept today in the University of Virginia, another of his big ideas). He studied it and called it a Megalonyx ("great claw"). Caspar Wistar (1761–1818), the anatomist who drew those claws and studied the case in more detail somewhat later, renamed it *Megalonix jeffersonii* in his honor (Figure 30).[33]

In February 1797, Jefferson traveled from Monticello to Philadelphia to assume the vice presidency of the United States (he had lost the contest for president by a small margin to John Adams). He took with him some of the bones of the Megalonyx to show them to the American Philosophical Society, of which he was to become president in March of that year. It must have been at about this time that he read Cuvier's article on the Megatherium in translation in the *Monthly Magazine*. What had seemed about to become another example of the ferocity of the American continent, a beast that was not the gigantic elephant of Ohio but something even bigger, its possible predator, an indomitable and colossal carnivore, was toppled from its pedestal in the twinkling of an eye. It did not take long for Jefferson and Wistar to perceive the similarities between the giant sloth of "Paraguay," our petrified skeleton in the Royal Cabinet in Madrid, and the discredited North American feline that had been unearthed, gloriously at first before disenchantment set in, beneath his beloved Virginia soil.

Jefferson was obliged to modify the article he was writing, which was eventually published along with Wistar's in 1799.[34] The huge creature on whom they had counted to redeem the dented pride of the New World was a near relative of a gigantic species, it was true, but it was midway between an armadillo and a sloth. The armadillo had personified the new continent in many images and maps from the late sixteenth century on, but the sloth was held in very low physical and moral esteem; indeed, it was the very animal that Buffon had singled out to illustrate the torpidity and apathy of American fauna. His description of the sloth is the most disparaging of all those contained in the forty-four volumes of his *Histoire naturelle*: "the most degraded animals [—] perhaps the only species to which Nature was unkind; they offer us a lively portrait of an innate disgrace ... with just one more defect, their existence would not be viable."[35]

IN THE MEANTIME, events and publications had moved ahead in Europe. Joseph Garriga, who had collaborated with Bru on other projects, published his report with the five plates in 1796 at his own expense after reading Cuvier's original article in the *Magasin Encyclopédique,* which he included in translation in his own publication. Why did Bru delay so long in publishing his work? Why did he not do so under his own name? The answer is unclear, but it is evident that he was spurred on by Cuvier's article. Cuvier mentioned the reconstruction and Bru's drawings (in fact, his article was based on that material and even reproduced the reconstructed skeleton), but apparently not sufficiently. Cuvier did note two or three gross errors committed by Bru (for example, the latter had confused the front and rear limbs and had referred to retractable claws like those of tigers), but without attributing too much importance to them. Quite simply, as an authority on vertebrate paleontology Bru was beneath Cuvier's dignity. If we believe what Garriga wrote in his prologue, his aim was simply to translate and publish the French naturalist's article together with the original plates, although afterwards, when he had read the general and particular description by Bru, scientific and patriotic zeal prompted him to convince Bru to let him publish it (in reality, to sell it to him).

In Madrid there was resentment toward the young Franco-German who had got in ahead of them and published an article mentioning the assembly and drawings by Bru in the Royal

Cabinet, but without doing them full justice. Iberian pride was wounded, Iberian science marginalized, or even worse, used and then superseded. Garriga went further and ventured to declare that Bru's description corrected the errors published by Cuvier in the *Magasin Encyclopédique*. His aim was to do justice to Bru and to the Spanish nation, whose naturalists "were not as sloppy" as the French suggested.

So the case was connected with the debate on Iberian science sparked by Masson de Morvilliers in his article on Spain in the *Encyclopédie méthodique* (1782), in which he posed the question of what the Spanish nation had done for Europe "in the last two, four, ten centuries."[36] The flexibility of the recently resuscitated vertebrate proved to be miraculous. It had just been at the point of being dragged into the cause of *libertas americana* as an example of the strength and bulk of the nature of the New World when it became a touchstone in the debate on Iberian science. Now, in the following century, the Megatherium would end up being enlisted under a different banner: that of its natural home, Argentina. It certainly was a polymorphous creature, a fresh and versatile cadaver, adaptable to a variety of political programs, patriotic interests, scientific theories, and credos, as we shall see.

There are still two shadows cast over Bru's description and plates that we can neither dispel nor ignore. Like the engravings themselves, the historical facts are full of gray tones and shades, doubts, fragments, speculations, the material of which its actors—and its authors and readers, we should add—are made. These are the accusation of plagiarism that Clavijo claimed to have heard from the surgeon of the Hospital del Buen Suceso; and the degree of participation by the surgeon, by Manuel Navarro, and by Bru himself in the drawings on which the engravings were based, drawings that are clearly different from others signed by the Valencian painter and taxidermist. They are reasonable doubts regarding a delicate question, authorship, especially if we bear in mind that Bru was embroiled in other episodes of the kind.

In fact, the questions concerning the authorship and property of discoveries (priority, plagiarism, and the like) are related to others that we have already come across in the history of the rhinoceros: the replicability of art and the mechanization of knowledge, the tension between false and authentic, copy and original. Ultimately it is a matter of the problem of the creation and ownership of a work. Who can claim ideas and scientific facts, the things we come to know, our own knowledge? Who is their author?

On the one hand, knowledge is a collective construction. There can be no doubt about its social character. In spite of the rhetoric of the first- or third-person singular with which the grand narrative of science has usually been told, as soon as any historical episode is considered more broadly, no matter how minor or insignificant it may be, we soon discover just how many hands and eyes contribute to the production of knowledge. On the other hand, science is too tied to the culture of modern art and the singular creative act to ignore them—the genius and the *eureka*, the crucial experiment and the *annus mirabilis*, the epiphany or revelation.

These are the borders of a problem that lies at the heart of the dialectic between science and art. "Art is I; science is we" is how it was formulated by Claude Bernard, the physiologist and father of experimental medicine. But this is to erect a boundary between public and private, reason and emotion, objective and subjective. For science is also I; art is also we. There is no sharp dividing line between the birth of the public and the emergence of the author—two constitutive processes of modernity—or between the history of science and of art. Rather, there is an ample shared space in which they employ similar resources, pursue related aims, and of course encounter very similar problems. In both fields, the pieces—the bones—are assembled and dismantled again, matched and articulated to fabricate images and stories that will explain or evoke the nature of things.

CHAPTER SIX

Fossil

> Genius and science have burst the limits of space, and obser-
> vations interpreted by reason have unveiled the mechanism
> of the world.
>
> Would there not also be some glory for man to know how
> to burst the limits of time?
>
> Georges Cuvier, *Recherches sur les ossemens*
> *fossiles de quadrupèdes*

During one of his visits to Paris in the late 1820s, the great
Scottish geologist Charles Lyell (1797–1875) managed to
gain access to the annex where Cuvier was working in the
library of the Anatomy Museum. Lyell must have been impressed, as he took careful note of the surroundings when he
entered the *sancta sanctorum*, a space that he called "truly
characteristic of the man." Cuvier's methodical style, the secret
of his splendid achievements, obtained "annually without ap-
pearing to give himself the least problem," could be observed
everywhere. His house was opposite the Museum of Natural
History, another building that he had reorganized, and was directly connected to the Anatomy Museum.

It was there that Cuvier had divided the library into a series
of adjacent rooms, each containing a different subject. Books on

osteology were open in one, works on ichthyology in another, while tomes on law were stacked in a third (it is not surprising that Lyell, a lawyer himself, should record this). Then Lyell reached the private study. He was shocked by the lack of bookcases. It was a very long room, comfortably furnished and lit, with eleven desks and two low tables, like a public office for so many clerks. "But all is for the one man," Lyell remarked in astonishment, "who multiplies himself as author" and does not normally allow anyone else inside. Cuvier moved around through that room as he passed from one task to another according to his needs and whims. Every table was supplied with its own set of inkwell, quills, and a little bell. The scholar used the low tables when he felt tired (he was about sixty years old). He had few assistants, but they were always carefully selected. They helped him to find references and spared him all kinds of routine work. They were rarely admitted to the interior of the study and generally confined themselves to receiving orders and never speaking.[1]

This description of the place where Cuvier worked was penned by the man who would shortly write one of the most influential scientific works of the nineteenth century, *Principles of Geology* (1830–1833), Darwin's bedside reading on board the *Beagle*. It is remarkable (and dispiriting) to see how time, the main protagonist in these pages, affects not only the course of nature on a large scale, but—more modestly but with equally visible results—the history of science too. It is surprising to what extent the limited time available to historians obliges them to lower the profiles of their protagonists and their works, leading us to underestimate the weight and influence of their ideas on their contemporaries or successors. Historiography is a capricious affair, sometimes rescuing minute facts and sometimes erasing (or at least blurring the outlines of) important biographies. A few years ago, during celebrations of the bicentenary of Darwin's birth in 1809 (and the century and a half since the publication of his *Origin of Species*), more than one scholar will have recalled the influence of Lyell, but it is probable that Cuvier was mentioned only to illustrate the ideas that were replaced by Lyell's geological uniformism and by the theory of evolution and natural selection in biology. Seen in this light, Cuvier appears as one of those scientists doomed to be associated with theories that are no longer upheld (catastrophism, the stability of species, creationism). The ideas themselves fall into disuse; unable to adapt, they become extinct, and fossilize. Historiography is not only capricious, but frequently opportunistic.

What is certain is that when Lyell visited Cuvier, the latter was at the peak of his fame; Darwin, by comparison, was then acquiring his first notions of natural history after abandoning his medical studies, to his father's disappointment. No student of anatomy and fossils was more admired and respected than the Frenchman, no one had such a network of correspondents, and certainly no one had so much power. Cuvier had succeeded Daubenton as professor of natural history in the Collège de France in 1799, the year in which he was also appointed executive secretary of the National Institute. He was later to become its permanent secretary, as well as presiding over the Imperial University and serving on the Council of State during the Napoleonic period. He continued to acquire positions in the Restoration: he was interim president of public education, was awarded the Legion of Honor, and somewhat later, in 1831, was knighted as a Peer of France. Shortly before his death, Baron de Cuvier was president of the Council of State and a minister.[2] By comparison, Darwin was at the time an obscure naturalist on board a ship bound for a remote destination.

Thus between 1798 and 1832 Cuvier wielded great institutional and scientific authority. He legislated and gave instructions in many fields beyond those he cultivated personally, in which he naturally exercised a strong hegemony. It is not for nothing that he has been called "the Napoleon of the natural sciences." Neither is it difficult to imagine him working simultaneously on various cases in the Museum of Natural History, as he did in the Anatomy Museum, comparing groups of bones of different species, observing their similarities and differences, their individual functions, or their influence on the skeleton as a whole. Methodical, with an attention to detail and sticking firmly to facts, Cuvier elevated the status of zoological osteology to an innovative scientific discipline with mathematical aspirations and applications in the field of geology and the history of the earth. After his identification—or rather invention—of the Megatherium, he managed to establish the profile of a whole class of fauna that had been invisible until then, a lost world to which he gave shape, thereby allowing his contemporaries and subsequent generations to imagine it.

The importance of Cuvier's work on anatomy can only be compared with the impact of his contributions on vertebrate paleontology and the history of the earth.[3] While in the field of comparative anatomy he postulated and applied two principles that were meant to raise the discipline to the level of the exact sciences (the subordination of characters and the correlation between parts), what interests us here is how he helped lead the way in an intellectual conquest with serious implications: to burst the limits of time, to take the expression that he himself used in the "Preliminary Discourse" to his *Recherches sur les ossemens fossiles de quadrupèdes* (1812) and that his most outstanding historian, Martin Rudwick, chose for the title of the first volume of his tour de force on the origins of geology and paleontology.[4]

One of Cuvier's most persistent ideas was the clear-cut distinction between extinct and living fauna, the abyss separating the present from a world that was relegated once and for all to the past. It is thus rather paradoxical that it was by applying the laws of anatomy to study both living and extinct species without distinction (a move already adumbrated in the act of naming an extinct animal with Linnaean nomenclature as *Megatherium fossile*) that Cuvier was able to demonstrate that both obeyed the same laws. But this is one of the ways in which science resembles art, as works in both fields move beyond the intentions of their authors and come to lead a life of their own.

By applying the same laws to the inhabitants of this lost world to those of the present one (something announced by the gesture of baptizing an extinct animal using Linnaean nomenclature), Cuvier was contributing to their unification, just as Newtonian physics had unified terrestrial and celestial bodies, culminating in the celestial mechanics that Pierre-Simon Laplace (a peer and major model for Cuvier) was conducting during the same fin de siècle. Cuvier dedicated his *Recherches sur les ossemens fossiles* to Monsieur le Marquis de la Place, the man who "had subjugated the heavens to geometry."[5]

Cuvier's work was a contribution to understanding and explaining the past and present of the earth and of life by drawing connections between them and introducing onto the scene what Foucault called "the irruptive violence of time."[6] Ideas about the great abyss of time were emerging from the depths of the earth, formulated by the Scottish mineralogist James Hutton (1726–1797) and now referred to as *deep time*, a term popularized by Stephen Jay Gould and the writer John

McPhee.[7] The notion is indebted to the Newtonian conception of infinite space, which included both linear and cyclical aspects of time, a dimension so vast that it went far off the human scale and, like everything else strange or unfamiliar, was difficult to recognize. One result of that deep time was what we might call the "alien past" (like naturalists, historians harbor the desire to give something a name): the past as a foreign country so old that there were not yet any human beings in it, an image that has contributed as few others have to put our species in its proper place. The alien past was considerably more distant for the Enlightenment than India had been for the Renaissance; less was known about it, so it was extremely difficult to imagine. If Ganda had transported that marvelous and imaginary Orient to Lisbon, the Megatherium was going to bring an infinitely more remote, prodigious, and strange territory in its wake.

The last decades of the eighteenth century were also those in which the last great voyages of circumnavigation sought to encircle and describe the globe definitively, and in which the universe appeared to be expanding indefinitely. Soon after James Cook and Louis Antoine de Bougainville completed the enterprise begun by Magellan and Elcano, Laplace concluded the vast Newtonian architecture, the cosmic vision offered by mathematics, physics, and astronomy, the disciplines that had dominated the Scientific Revolution—as traditionally considered—thanks to which space had been explored. The turn had come for the revolution of time at the hands of the life sciences. In comparison with what this revolution was going to entail—the grand history of the earth and of living creatures, the real history of the human species—the other revolutionary events (the Directory, the French National Convention, the Napoleonic Wars) become mere anecdotes.

ONCE THE EARTH had been circumnavigated, the secrets of its interior had to be revealed. Once the surface had been surveyed, or at least the main coastal profiles, work could begin on excavating and exploring its depths. The creatures and natural products that began to emerge from its interior prompted the formulation of new texts and scientific theories. As in the study of human anatomy, in which surveys of the external forms had been succeeded by dissection and practices that brought to light the hidden viscera and organs (including the bones), something similar was done to the earth.

What came to light were fossils that, as the Latin root *fodere* (to excavate) indicates, had been dug up from the earth. They were not new; indeed they had been known since antiquity. The history of how they had been understood and of the ends to which their study had been put is a long and fascinating one. We should recall some of the precedents to understand that, although the Megatherium itself was new (the newest kind of animal at the time because, paradoxically, it was older than the hermetic texts were for the humanists), it did not emerge out of the blue. Unlike the ancient portents, those divine signs that warned of future events, this one came from the interior of the earth, and what it had to say concerned not the future but the past.

It was not a preternatural being, a phenomenon that could link terrestrial events with what lay beyond the earthly sphere.[8] It may have had something in common with a portent as its teratological condition never left it, but there could be no doubt that it was a fossil—in other words, a decidedly subterranean creature. As Foucault put it, both monsters and fossils posed problems for taxonomy, neither of them fitting easily into contemporary schema, into the various epistemological arrangements that scholars typically delineated with tables and continuums.[9]

Our creature was a fossil—before its identification, this had been one of its three recognizable features, as before becoming a Megatherium it had been the bundle of fossilized bones of a monster. It had been unearthed in an alluvial basin that favored sediments of this kind. As a fossil, it was the organic remains of a creature that had been petrified by time, an organic product that had been mineralized. Midway between a stone and a living being, between the inorganic and the organic, this disconcerting, almost magical dual nature was one of the most striking characteristics of fossils. Composed of diverse and in a certain way opposite elements, fossils shared with chimeras the quality of being hybrids.

The category of fossil covered a wide range of products whose nature was difficult to determine. Its origins and properties had been debated since antiquity. It was uncertain where it came from or how it was formed. The Aristotelian tradition credited "vaporous exhalations," an idea that two philosophers, the Persian Avicenna and the German Albert of Saxony, transformed into the hypothesis of a fluid that was solidified or petrifying (*succus lapidificatus*).[10]

An agent or a process that petrified organisms? In the dawn of the early modern period, a fossil was much more and very different from what we understand by the term today. It was a petrifaction, a stone in a much wider sense of the word. It was

not easy to know what it was, or when a stony or rocky formation was or ceased to be one. Fossils had a material consistency
but an organic appearance in the form of a plant or animal. But
did the fact that they had the form of living beings really mean
that they were living beings themselves? Or had they once been
living beings before losing their organic nature? There were organized fossils, strange fossils, apparent fossils, fossils in the
shape of a star, an ear, or a heart, while others could only be a
tooth, a shell, or a petrified leaf. All of them—ammonites, gems,
precious stones, crystals, whale ribs, meteors, the maxillaries of
a large quadruped—were studied by the same individuals,
appeared in the same collections, and were reproduced in the
same treatises.

Conrad Gessner (1516–1565) deserves a place of his own in
this chapter of the history of science. Author of the magnificent
Historia animalium (1551–1558)—one of the channels through
which Dürer's image of the rhinoceros was disseminated—
Gessner also wrote a text that is a forerunner of modern paleontology, *De rerum fossilium, lapidum et gemmarum maxime, fig-
uris et similitudinibus liber* (1565). It was conceived as an appendix
to his grand project of an encyclopedia of natural history and as
a preliminary essay to a more extensive treatise on fossils that he
never managed to write. Although intended as an introduction,
it was as rabidly erudite and philological as the rest of his work.
And, like his *Historia animalium*, it included images, which was
a genuine novelty. The woodcut illustrations, like those that Leonhart Fuchs had printed in his botanical treatise *Historia Stirpium* (1542), played a fundamental role in the communication of
knowledge in general terms, as we have already seen with Dürer's engraving. In this case, given the primarily descriptive character of knowledge about fossils, the illustrations were much

more than a mere accompaniment or ornament for the written text.[11]

In the face of the diverse names that fossils had been given in different languages and the sprawling proliferation of heterogeneous information that the humanists liked to collect, what was needed to advance discussion was to make that information homogeneous and to present the facts themselves or, failing that, the images that stood in their place. The spread of printed images was the only way to overcome the problems of long-distance communication, terminological variation, and the very limits of language, especially when dealing with mute, bare facts like these. Fossils emerged from the depths of the earth like surprise witnesses in court. They were here. They had no answers. They were surprising and wonderful.

In addition to harmonizing how they referred to fossils, adherents of the Neo-Platonist and hermetic traditions in Gessner's day were interested in fossils as analogies of other phenomena and even for their alleged therapeutic qualities. Like copies or imitations, fossils were duplicates, correspondences, images of other living beings—a recurring subject in this essay—that had echoes in Plato's cave. What were they? Had they ever been animate? Or did they simply look that way? The theme of hidden harmonies and resonances in natural objects intersects with the (problematic) relations between art and nature, whose variants modern man has explored to exhaustion: the metaphor of the world as an artifice is an idea, a cultural and aesthetic referent, that pervades the science, poetry, theater, and painting of the modern age. From this point of view, fossils were *lusus naturae*, natural artifices, nature's playthings or diversions, and demonstrations of its capacity to model. If people imitated the Creator with their art and science (penetrating and obeying nature, as Bacon wanted), it was hardly strange that nature herself should imitate and reproduce her own work in this ingenious way. Moreover, adding to this analogy with art, fossils were easy to preserve and display. They seemed designed to be stored and exhibited in a museum.

As long as completely satisfactory explanations failed to materialize, those who took an interest in fossils could only accumulate and exhibit them, reproduce and share them, catalogue and sketch them, and verify how many similar or identical pieces had been found in other latitudes. Before 1600 there were already significant collections of fossils and minerals in the Vatican, Bologna, Verona, and Naples. They were typical objects to be found in the *Kunst- und Wunderkammern* and *metallothecae*, those collection spaces of what Giuseppe Olmi has called "the *longue durée* of curiosity" (Figure 31).[12] Catalogues of the collections' pieces and treasures were published; the illustrations were the primary vehicle for conveying knowledge in this broadly collaborative enterprise. The need to share information, to communicate finds over distances, and to compare cases made the study of fossils a visual field of knowledge pursued collectively, what in the Internet era would be called science on the web.

Another name that cannot be left out of the prehistory of geology and the study of fossils is Nicolas Steno (1638–1686), a Dane employed by the Duke of Tuscany in Florence who underwent more than one conversion in his life. Born the son of a Lutheran pastor, he eventually became a Catholic bishop and a missionary on orders from Rome, abandoning extreme Protestantism for militant Catholicism. He likewise abandoned anatomy and medicine for the study of rocks and fossils. In fact, as often happens, he shifted his gaze from one camp to

Figure 31. Detail of one of the earliest images of a cabinet of curiosities, engraving in Ferrante Imperato, *Dell'Historia Naturale* (1599).

another—a polite way of saying that he was buffeted from one problem to another. After spending years on the similarities between human and zoological anatomy (comparative anatomy is as old as Aristotle; in the last resort, all men of science study what is identical, similar, and different), he came across the strange case of the *glossopetrae* (stone tongues) and their intriguing resemblance to sharks' teeth. Applying a mechanistic framework, Steno narrowed his interest to the attempt to explain how one solid had been formed within another solid, as the full title of his book phrases it: *De solido intra solidum naturaliter contento dissertationis prodromus* (1669).[13] As with Gessner's *De rerum fossilium* and many other cases in the history of science, this was the introduction to a prospective work that its author never managed to write. Instead of basing his study on the formal appearance of fossils, Steno classified them according to how they had originated, that is, how some solid bodies had managed to become lodged inside others. He distinguished clearly between rocks or crystals formed in the interior of the earth and organic fossils. He enunciated the principles of molding and sufficient similarity and established the rudiments of stratigraphy, thereby clearing the way for the chemical study of fossils and their stratigraphic record. These were the first steps in what we now call crystallography, and in historical geology the study of the origins and past of the earth. These were matters that for centuries had been studied in the Western tradition in light of the biblical account. This does not make them prescientific or even nonscientific, for only an excess of presentism or a serious lack of information could lead us to expect the possibility of a separation between the knowledge of nature and that of providence at this time.

Robert Hooke (1635–1703), England's Leonardo, was also interested in fossils.[14] The man responsible for the experiments of the Royal Society wanted to know how the petrifying liquid permeated the bodies; why ammonites looked like certain pearl shells of the East Indies, the nautilus; and how remains of animals that lived in the tropics or in the sea could be found in northern latitudes or inland (Figure 32). Following the Neo-Platonist tradition, Hooke vaunted the capacity of nature to

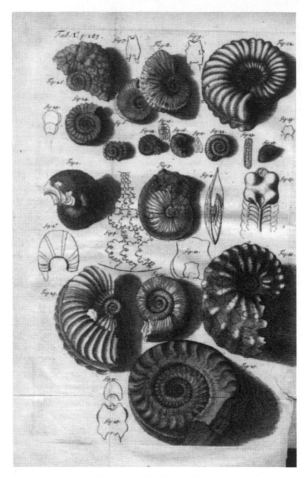

Figure 32. Ammonites drawn by Robert Hooke and described in his *Discourses of Earthquakes and Subterraneous Eruptions*, published after his death in 1705.

mold. In this he differed little from a man like Athanasius Kircher, for example, that remarkable figure who seems to have emerged from a short story by Jorge Luis Borges. There must have been some kind of *spiritus plasticus*, he wrote, a geocosmic liquid capable of permeating bodies, which then became petrified as they grew moist, congealed, or became fused, one of those explanations that can be ridiculed for being no less fantastic than those it sought to displace.

Still, Hooke's response was an original one. Like Steno, he sought the explanation in the interior of fossils. Unlike the Dane, however, he used a microscope, the powerful instrument by which realistic images could be obtained by the simple combination of "a faithful Hand" and "a sincere Eye." So he focused his lens on some fossil wood remains and discovered that its structure closely resembled that of carbonized wood. Hooke verified the penetration of petrifying liquid to the interior of the organisms and defended the organic origin of fossils in his *Lectures and Discourses on Earthquakes* of 1668, though these were not published until after his death. Hooke speculated with great acuity on the nature and origin of fossils and the bones of certain animals that did not seem to match those of any existing animal. He was the first to apply the microscope to the study of fossils and postulated an idea that was truly bold at the time: the possible extinction of living forms in the past. He adduced the movement of mountain ranges and changes in the sea level to account for their present location. He even compared fossils with medals and ancient coins, another fruitful idea.[15] Hooke clearly linked stratification with the age of the remains and can be considered as much a visionary precursor in geology and paleontology as he is in so many other fields.

The organic origin of fossils was denied by Martin Lister (ca. 1638–1712), a physician and scrupulous student of British shells. He was joined in this debate with Hooke by John Ray (1627–1705). An extremely influential naturalist in his day and in the British Isles until the appearance of Darwin, Ray had begun as a student of the local flora of Cambridge and ended up becoming one of the major representatives of natural theology and the argument from design. He admitted that the petrifactions might be more than copies of living organisms, but refused to countenance the possibility of extinction because it contradicted the providentialist vision and the principle of plenitude. The species were fixed and immutable. God had created the variety of species in their present and recognizable form. If there were fossils whose living counterparts were unknown (the proof of extinction), it was simply because they had not yet been discovered: they must be hidden in some coral reef of an archipelago in the Indian or Pacific Ocean.

At any rate, knowledge about fossils had increased considerably by the beginning of the eighteenth century. Collections had multiplied, the doctrine that nature was organized like a machine was gaining ground, and the new techniques of observation and the reproduction of images were improving naturalists' descriptions and analyses. More information was circulating more effectively. Fossils were a recurring subject in the sessions of the new scientific academies, and it was rare for an issue of any of the new periodicals (scientific journals had just begun being published) not to include an article, description, or commentary on the discovery of a remain or specimen in some cabinet or other.

Fossils would soon become a crucial point when it came to adjusting two books, the book of nature and the book of the scriptures, the two divine inscriptions in the world that marked the horizon of all science at the time. As we have seen in the history of the rhinoceros, natural phenomena had to be accommodated to the written word, and above all else, to the written "Word of God." Some historians treat this with a measure of condescension, as if their method (or ours, to be honest) were so different, for what do we do if not hold up the events of the past against the tables of the law of the present and our own ideas about how wrong Cuvier was and how right Darwin was?

We treat the objects of the past in the same way that paleontologists scrutinize the courses of past lives. As if they were fossils, we delight in finding similarities, precedents, analogous forms. We are always ready to recognize what is familiar or proximate, but how is one to recognize what is really new or what once existed but no longer does? How are we to distinguish what is unprecedented, unfamiliar, irrevocably extinct? How can we imagine the unimaginable? The history of the Megatherium is situated in a fold of the past in which things unheard of were beginning to be conceived.

THE POINT OF intersection at which fossils were situated between the Book of Nature and the Holy Scriptures was also the point of intersection between two histories, human and sacred history, whose fundamental episode in geological terms was the Flood.[16] Nature was a book, but it did not have a history. It was like a text in need of deciphering and interpretation: that is how it was understood thanks to the power of the printed book, the instrument that, by exercising its hegemony in the world of knowledge, ended up projecting its image onto the world itself. But it did not have a past, or at least not a past as we understand it today. Natural history was the complete opposite of the history of nature; throughout almost the entire early modern period, the study of fossils took place within a descriptive, taxonomic, profoundly ahistorical discipline. All that was known about the history of the earth was shrouded in the night of time and interpreted to conform with the biblical narrative, beyond the reach of the sources of moral and civil (that is, human) history.

What was known about the rhinoceros (and by extension eastern fauna) in the Renaissance was derived from classical natural history and the kinds of legends and fables that hover around frontiers. The little that was known about the origins and past of the earth at the beginning of the eighteenth century was illuminated by the book of Genesis. Certainly the observations and experiments of modern science would end up by reducing Genesis to the status of an allegory or myth about the origin of the world and of the human race, but it would be wrong to consider that the Bible prevented the emergence of a notion of time that would be useful for thinking about the earth in historical terms. In fact, exactly the opposite occurred: just as Pliny's allegorical ancient natural history and orientalizing legends had shaped the figure of the rhinoceros, Genesis modeled the first histories of the earth to such an extent that it inspired a genre of history that came to be called sacred physics.[17]

Hebrew history provided the framework of a dramatic, meaningful, and climactic narrative. Not only was it sanctioned by religious orthodoxy, but it was also more plausible (for its realism) than the tall stories of the Greeks, more linear and less perpetualist than the Oriental legends of Aristotelian philosophy, and less cyclical and all-embracing than the hermetic or Neo-Platonist traditions.[18] The stages and levels of Genesis, from "Let there be light" to the division of the waters and the creation of dry land, culminating in the making of man in God's image to crown the world of living creatures, constituted a model that made it possible to assign a place in history not only to human events (of course, Jewish history lies at the foundation of Christian historiography) but also to natural events. What is more, it enabled both to be seen as part of the same plan or narrative. We find this idea here in an embryonic form, even though the Bible and its exegesis set man above the rest of the creation (this exegesis stretching from the Christian Fathers and Saint Augustine to John Ray and Buffon, via Aristotle, Descartes, and many other pillars of Western philosophy and science who erected an insuperable barrier between the rational animal and the rest).

How old was the earth? No instrument was capable of measuring such a distance, no lens able to reach so far. The past cannot be experienced or observed directly; it has to be reconstructed or imagined on the basis of remains and traces, ruins, fossils. Given the absence of instruments for visualizing the past, chronology was probably the tool of calculation. Based on philology and astronomy, chronology is one of those disciplines that we tend to look down on today, but it occupied the brightest minds of the seventeenth century. Biblical chronology implied that the earth could not be more than six thousand years old.[19] We may smile today at Archbishop James Ussher, who calculated the date on which God had created the world with the precision of a clockmaker—Sunday, October 23, 4004 BC—but it is too often forgotten that even Newton took a greater interest in chronology and biblical exegesis than in the laws of motion.

The remains of elephants in Italy must have arrived during Hannibal's campaigns; the petrified shells to be found on the highest peaks must have been carried there by the Flood. Some of the geographical shifts and changes that could apparently be deduced from observations were attributed by Hooke to earthquakes. The earth had changed over the years. Like a living being, it must be growing, or was it growing old? Whether optimistic or pessimistic,[20] such visions of nature touched on a delicate and fundamental question: Was one to accept change and alteration and to recognize the effect of time on life (or whatever life was called at the time—if we follow Foucault in observing that life was inconceivable from within natural history)?[21] Indeed, the mutability of nature, the matter that hinges on the question of fossils and changes in the earth, is a theme that goes far beyond the boundaries of natural history and permeates many other cultural phenomena of the early modern era (still-life painting, religious literature, the dramas of the Spanish Golden Age).

Among the arguments regarding the changes that had taken place in the earth, the hypothesis of giants gained increasing support from the end of the seventeenth century. Robert Plot, curator at the Ashmolean Museum in Oxford, came across a distal fragment of a gigantic femur. He first thought it belonged to an elephant that the Romans had brought, but went on to assign it to some antediluvian patriarch, one of those giants who had walked the earth before the Flood.[22] He had a drawing and an engraving made of it. When the encyclopedist Jean-Baptiste Robinet contemplated the same illustration a century later, he identified it as a human scrotum *(scrotus humanus)!*[23] The arrival of elephants in the British Isles was documented; that people had been larger in the past was a widespread notion that found confirmation in numerous remains in different parts of the world. Either of the two hypotheses was scientifically more probable at the time than what we know today to have been the case: the femur was that of a Megalosaurus, a dinosaur, an animal as unimaginable for Plot or Robinet as the Megatherium had been for Manuel Torres or Juan Bautista Bru.

The age of the earth implied the age of humanity. What was its past before what the sources and inscriptions told us? Humankind's roots must have gone back approximately to the time of the earth's; it was difficult to imagine anything else. Like children who are unable to imagine the life of their parents before they were born, it has taken us (at least those of us in the West) a good time to imagine the life of the earth without our presence, let alone reconstruct it. The alien past was so foreign that it was difficult enough just to conceive it.

In the meantime, however, astronomy was expanding the limits of the universe, shifting our planet from the center and placing it in a minute place amid a literally infinite space. Descartes and Leibniz drew far-reaching philosophical conclusions and made major contributions to a genre of thought that was gradually becoming more physical than sacred—in other words, that allowed a certain autonomy to physical facts on the margin of providence. As in everything, there were degrees. Divine action was more constant and visible in some hypotheses than in others. We should not draw too many conclusions from our secularism, as often happens: Leibniz and Newton should remind us of the presence of numerous theodicies and of the fact that some of the most sophisticated scientific developments were assisted rather than thwarted by religion.

The Cambridge Neo-Platonists also worked to reconcile biblical chronology with the new conceptions of the universe. They set their mark on the theoretical horizon of Thomas Burnet (ca. 1635–1715), another essential author for us here, however summary the account. His *Telluris Theoria Sacra* of 1681, published in English three years later as *Sacred Theory of the Earth,* fell back on a tradition containing many of Steno's ideas but spruced up by the theories of Henry More, a bit of Descartes, and even Kircher. Perhaps Kircher was one of the last of the men who knew everything available to know; and certainly he was one of the first to venture to represent the interior of the earth (another territory as unknown and impenetrable as the past, which could only be imagined but never seen) in his *Mundus subterraneus.*[24] Kircher assigned a key role to volcanic activity in the configuration of the surface of the earth (what would later be called Plutonism), but Burnet regarded the Flood as the prime cause of change on the earth's surface. His reconstruction of the origins of the world was spectacular and captivated the imagination of his epoch.

He based his theory on all kinds of scientific evidence. He used and confirmed the biblical chronology, though he took it allegorically rather than literally. Adopting a generalized deism laced with prophecies of the millennium, Burnet described Paradise and predicted the coming of Christ and the New Kingdom of Heaven. He reconstructed the seven ages into which history could be divided: the past but also the future, when the earth would be converted into a star and would be consumed in its own fire (Figure 33). This cosmic and highly poetic vision was to inspire Coleridge decades later. The earth had been created completely flat, without peaks, valleys, or rivers. It was formless and lifeless, neither hot nor cold. There were no tempests or storms. The temperate days had a longer duration at that time (this greatly pleased Newton, who was concerned with the rotation of the different spheroids). Then came the Flood and chaos. Mountains emerged, God caused

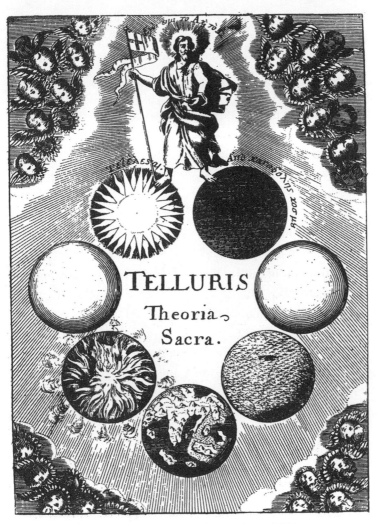

Figure 33. Thomas Burnet, *Telluris Theoria Sacra* (1681).

the waters to move, but as they were not sufficient, he had to contract the earth (which explained the shorter durations of the days on the edge of its internal structure). This contraction of the globe produced the reliefs and chains of mountains, the coasts and cliffs. It also triggered the succession of the seasons, imbalances, and other changes in nature.[25]

It was a catastrophic, redemptionist, tragic vision that was common in the age of the English Revolution. The world was growing old, wearing out like a papyrus, in ruins.[26] Burnet did not concern himself with fossils, but the *Essay toward a Natural History of the Earth* of 1695 by John Woodward (1665–1728) did and was deeply influenced by Burnet's images. So was Johann Jakob Scheuchzer (1672–1733), the Swiss author of another fantastic work—a treatise on the six days of the Creation—halfway between epic literature and geohistory.[27] Scheuchzer identified the skeleton of an amphibian from the Cenozoic era with a human witness of the Flood *(homo diluvii testis)*. Some of these ideas were wide of the mark, others had a better foundation, and some were both (there are many well-founded ideas in outlandish arguments, however strange and far-fetched they may be, not to mention the number of outlandish ideas that have been well argued). But somewhere in their confluence the idea of the organic origin of fossils, their relation to the Flood and to the history of the earth, began to be consolidated.[28] Evidence was piling up that confirmed the planet had undergone changes to its surface. These changes (volcanic action, emergent mountains, waters that had flooded and displaced large terrestrial masses) had scattered stones and fossils on the surface in places far from where they had originated.

The eighteenth century witnessed the emergence of grand theories of the earth and of fossils. As in so many respects,

Leibniz (1646–1716) was a chapter by himself. His *Protogaea*, posthumously published in 1749,[29] argued that fire was at the origin of all change. The earth had been an incandescent mass (a Cartesian idea), although fossils, he said, were the result of the progressive deposits of an ocean that originally covered the entire surface. Such Vulcanist (or Plutonist) and Neptunist theories vied with one another even before they had been formulated as such. The experts had to choose: Which had been the bigger cataclysm, the Flood or the volcanic eruptions? Which had come first? Where did the first stones come from? Had water preceded fire, or the other way around? Leibniz combined the ideas of Steno and Woodward with the sacred scriptures and the new science. He appealed to the condensation of liquids and envisaged a succession of floods instead of one big cataclysm. He also found it more plausible to suppose that the Creation had been not one act but several. With characteristic lucidity, he considered the possibility that some species might have been transformed into others. Nobody yet suspected where the old idea of the Great Chain of Being was going to lead.[30]

Neither did the Comte de Buffon, a boundless personality and one of the last great naturalists of the Republic of Letters. The last director of the Jardin du Roi was a literary talent who moved in aristocratic circles, a naturalist philosopher, one of an almost extinct breed.[31] In spite of the attempts by some historians to present him as a precursor of Jean-Baptiste Lamarck, today we know that Buffon always believed that the species were stable and constant. He only accepted degeneration (transformism) within the same species and on an individual scale. All the same, his belief in fixity, which ruled out evolution, did not stop him from being one of the first to maintain that human history was only a brief episode in the history of the earth (though Buffon could not have suspected how brief—barely the twinkling of an eye—that fleeting episode was). In Hooke's day the history of the earth was calculated in thousands of years; now people had to get used to the idea of millions or even billions, something beyond the ordinary experience of any human being and thus extraordinarily difficult to conceive or to imagine.[32]

In his work *Les époques de la nature* (1778), Buffon took up Leibniz's idea of an originally incandescent earth that gradually cooled down over the centuries. Like the seven days of Genesis, he envisaged seven stages in the history of the globe, but unlike Burnet, he relegated all of them to the past. Humankind, the king of creation, a being completely different from and even opposed to nature, made an appearance in the last of these (Buffon had an elevated concept of humanity, or at least of a part of it, to be precise). He also drew on fossils to reconstruct that long history over tens of thousands of years, perhaps millions. He demonstrated the differences between mountains of primary and secondary deposits, the accumulations of rocks and fossils before and after living beings. The size and location of some of them (ammonites and some remains of large northern elephants) confirmed the tropical climate that predominated in a past that was expanding with giant steps.

It was a question of scale. The Scot James Hutton, whom we met earlier, is rightly considered one of the fathers of modern geology, a pioneer of the idea of deep time. Hutton studied the processes of erosion and sedimentation in detail, as well as the changes in the angle of different deposits (the latest formations rest on older ones at a different angle). Against theories claiming that rock formations were created by inundation or the effects of the oceans, Hutton defended the Plutonist theory, according to which they were produced by heat from the fiery interior of the earth. After that, erosion by air and water took place. The materials were sedimented in layers until they finally took the form of solid rock, due again to the effect of heat. The earth was not irrevocably growing old; it was renewing itself. Hutton also had the bright idea of supposing that the past had not been so different from the present (actualism or uniformism), so that the action of these agents had been very gradual and the processes necessarily extraordinarily long. The earth must be more than a few thousand years old. Its history required a super-human calculus, a different, more grandiose chronology: geological time. In fact, the earth became a machine without history, caught up in a cyclical time without a trace of beginning or signs of end.[33] Hutton's notion of indefinite time was inspired by Newton's ideas on space.

But this notion of deep time did not emerge just by contemplating the cosmic dimensions of the heavenly mechanism, or the theories of the earth reviewed above, or even those of the long-term processes of rise and fall that characterized philosophical history à la Voltaire or the grand panoramas of such historians as Edward Gibbon, William Robertson, and Giambattista Vico. The changes did not come from such noble quarters, although contemporary histories provided the new scientific ideas with points of reference, some because they were the traditional source of ideas about the deep past, others because they threw light on historical processes or phenomena that might resemble the history of the earth. The changes, rather, came from below, from more humble origins. It is common for historians to grant new ways of doing science a noble pedigree and an exalted intellectual family tree, what we might call an upwardly mobile and mythologized genealogy of the new, rather as Lorenzo de Medici presented himself as a latter-day Aeneas or Caesar. What is certain is that many of the most important changes in the forms of producing and communicating knowledge have often come from more humble, artisanal, plebeian origins. Without Neo-Platonism, Botticelli would never have painted *Venus;* without Newton's physics there would have been no Enlightenment. But without the printing press, engraving, or the manufacture of instruments, neither Neo-Platonism nor Newtonism would have become canonical.

IN THE YEARS leading up to the French Revolution a broad consensus was emerging about two principal matters: that the history of the earth must go back much, much further than had traditionally been supposed, and that fossils could be more useful for the study of that past than had been supposed also. Almost nobody was prepared to abandon the biblical narrative, but there were many ways of interpreting it. No naturalist worthy of the name took it literally, though this did not prevent most of them from giving an important role to the Flood in the configuration of the earth's surface. On the eve of the taking of the Bastille, the Flood was still the major geological agent under another name. In any case, sacred physics had been considerably naturalized. A Euhemerist wave (for Euhemero, that demystifier of the Greek gods so beloved by Voltaire) was sweeping through the Enlightenment as Voltaire and his supporters allegorized the gods of the Greeks, just as the biblical account was coming to be rationalized or naturalized too. This rationalization may have had the effect of confirming the Flood, but it also generated an accumulation of knowledge that could function autonomously from its biblical origins when the time came. Mineralogy and chemistry were making such progress that a new discipline, geology, finally emerged from the study of relief, rocks, fossils, and basalt (the latter was very hotly debated by the Vulcanists and their opponents). Henri Louis Frédéric de Saussure prepared to climb Mont Blanc in 1786, and Hutton was soon to write the first version of his *Theory of the Earth* (1788). The champion of Neptunism, Abraham Gottlob Werner (1749–1871), a professor at the influential Mining Academy in Freiburg, had published a big monograph on minerals and fossils, *Von den äusserlichen Kennzeichen der Fossilien*, in 1774.[34] Fossils were still studied together with minerals. Neither lithology nor paleontology had yet been formalized.

On the other hand, with the bursting of the limits of time and the incipient realization that the history of the earth and of some living creatures had begun long before the history of humankind, fossils ceased to be illustrations of or supplements to human history. They were the only evidence of that remote, prehuman history. The analogy that Hooke had drawn between fossils and ancient coins or medals grew and developed a century later, though in directions unforeseen by our courageous microscopist (Figure 34).

In 1784, Tylers Museum in Haarlem, Holland, held a competition for the best dissertation on the relation between fossils and the changes in the surface of the earth. It was won by François Xavier Burtin (1748–1818) for a work on "the revolutions in the surface of the earth and the age of our globe." Burtin, who had studied in Louvain, had already established similarities between the nautilus shells of the tropical oceans and the petrified remains found in the vicinity of Brussels. The idea that fossils were like coins, ancient monuments, or documents of

Figure 34. Coins and fossils. The usual analogy between these two objects considered both to be traces or remains of (social or natural) past forms.

nature was now taken seriously. They had to be read and deciphered. The study of their forms, composition, stratigraphic location, and, as we would say today, "the context of the discovery" were ways of gaining access to the history of the earth. A naturalist now had to become a historian of nature, a researcher on soils, minerals, and fossils (a geologist), a historian of the earth. Mineralogy was an instrument for historical geology. From this perspective, Burtin confirmed what had long been suspected: the difference of scale between human history and the history of the earth. In his opinion, the great revolution of the globe could not have been the Flood, but a much bigger and far older phenomenon. As he refined his ideas, he deduced that the Flood had not been the only catastrophe or crisis of nature.[35]

Jean André Deluc (1727–1817), another pioneer in geology and the first to use the term, disagreed. Like Scheuchzer and Saussure, Deluc was Swiss (the Alps inspired not only many artists like Dürer but also many geologists). In his experiments, he had perfected the instruments to measure temperature and humidity, and deduced certain laws from their behavior. But he was also a confessed Neptunist, who desperately tried to fit the biblical accounts into the long history that seemed to be taking over soils and fossils. A decade before Cuvier began to study the Megatherium, Deluc was defending the importance of the Flood in the geological history of the earth. He characterized it as a catastrophe, a major revolution that had transformed the world from top to bottom. Contrary to the uniformism that Hutton postulated, in Deluc's opinion the Flood marked an abrupt fracture between before and after. History could be read in the inscriptions and ruins of nature (in the mountains and volcanoes, in the strata and sediments)—but it was a history that spoke of two different worlds, one that was irrevocably lost and one that was relatively recent.[36]

But how different was the lost world? What did "revolution" mean at that time and in this type of debate? What was a catastrophe? Where did all that talk about ruins and inscriptions, medals, and coins of nature come from? Deluc claimed that the two worlds were clearly different, because he understood the term "revolution" not in the classical astronomical sense as the planetary revolutions, the orbits that they described around the sun, but in the sense that is more familiar to us today: as a crisis. Johann Friedrich Blumenbach (1752–1840)— better known for his anthropological investigations of race— would popularize this notion in speaking openly of a *Totalrevolution* in an article significantly entitled "Naturgeschichte der Vorwelt" in 1790.[37] Blumenbach explicitly argued the case for the existence of a world before Adam that had disappeared completely; he denied that differences between present-day and primitive (fossilized) species could be explained by any kind of transformism or gradual degeneration over the course of time, not even millennia. A rift had emerged between the first and the second world as a result of some major natural catastrophe. One had collapsed; the other had emerged (or had been created) entirely afresh.

Seen in this light, fossils had to be studied as (the only) traces of that lost world, just as coins, medals, and ceramics had been used to reconstruct antiquity. The language and analogies of the antiquarians and the history of the scholars burst onto the scene in the final decades of the century.[38] The ruins of Herculaneum (1738) and Pompeii (1748), both discovered during the reign of Charles VII of Naples (later Charles III of Spain), were providing inspiration beyond the discipline of classical archaeology.[39] The discoveries were to provide a definitive stimulus to archaeological studies more generally and to shape a model in which the history of the earth itself appeared. This was because, like the founders of geology and the study of the history of the earth, the precursors of archaeology worked with material remains. Although the latter had certain written texts, inscriptions, and documents at their disposal, the founders of the history of the earth did not. In its more extended form, history—which had been compressed for centuries under the

unquestionable hegemony of the word and the written document—was at last incorporating material evidence, artifacts, and
objects as sources (even as autonomous and sufficient sources)
for the reconstruction of the past.

The eruption of Vesuvius had "frozen" the life of cities, providing a snapshot, as it were, of a discrete moment. Domestic
life, temple rites, articles of trade, even the bodies of the inhabitants were buried beneath the ashes, carbonized and immortalized, in a sense, like fossils. It is difficult to exaggerate the
impact produced by that vision on the imagination of the *savants* and *philosophes*. The Scottish diplomat, antiquarian, and
vulcanologist William Hamilton (1730–1803) was one of the
main figures who popularized the discoveries. Years earlier, the
Aragonese civil engineer Roque Joaquín de Alcubierre (1702–
1780) had conducted the excavations that brought to light images of the world of antiquity—images that were infinitely more
expressive and eloquent about the texture of Roman life than
all of the existing literature and book knowledge combined.[40]
Thanks to the archaeological works in the surroundings of Naples, from the middle to the end of the century educated
Europe witnessed the resurrection of ancient Rome, the culture that affected political theory, historiography, and so many
branches of knowledge during the Enlightenment. The images
and objects overshadowed words in an activity that was rapidly
understood as an epistemological example to be reckoned with
(Figure 35). And that was not all: that dramatic episode offered
a paradigmatic case of the intersection between human history
and the unleashing of the forces of nature. A natural catastrophe caused one civilization to collapse, and now, centuries
later, its petrified ruins made it possible to recover a lost world.
Did the meticulous students of fossils need any more clues?

Figure 35. "Veduta interior o atrio d'un Tempio nella parte occidentale di Pozzuolo," in Paolo Antonio Paoli, *Antichità di Pozzuoli* (Naples, 1768).

Not many more, for sure. Installed in the tradition of the grand theories of the earth although mildly skeptical of their speculative character, inclined as they were to follow the example of more empirical and modest scientific practices such as mineralogy or the study of antiquities, the naturalists took up the challenge to mount a definitive assault on the mysteries of fossils, those millennial secrets of nature.

Deep time was taking shape in the imagination as a result of the cyclical and linear conceptions of an ensemble of branches of knowledge that dealt with the matter between 1680 and 1830, between Burnet and Lyell, between mosaic and biblical exegesis and the rise of mineralogy or the study of antiquities.[41] When Cuvier published his first work on the Megatherium in 1796, there were two main responses to the basic question regarding the past of the earth: the gradualist, uniformist model of Hutton, and the catastrophist model of Deluc. Besides the role that they assigned to fire and water in the origin of minerals, the two models differed on how the changes had been produced. The first held that they were the result of continual erosion, sedimentation, and heat in the course of millions of years, a deep time that was expanding indefinitely. The second attributed everything to the violent outburst of a catastrophe—the Flood or some other revolution of nature. According to the first, the changes continued to occur into the present: the present world was governed by the same laws as the old one. According to the second, the changes had been produced in a moment of crisis, preceded and followed by highly stable epochs, with the outcome reflected in two completely different worlds, the lost world and the relatively recent one.

The fossils accumulated in the cabinets and museums of Enlightenment Paris (no doubt the most well endowed city in this respect) could be used to back up either theory.[42] When treated like coins or temporal markers of the living creatures of the past, just as volcanoes or mountain chains were markets of the earth's past, many fossils seemed to indicate the existence of a completely different fauna from what was known and alive. This lent support to the theory of catastrophe and extinction. However there were two other possibilities—migration and transmutation—which were in a certain sense more probable than the first.[43]

Extinction called into question the role of providence. How was it possible that God had destroyed his own work? In the story of the Flood, Noah had taken care to preserve the species in the ark. The hypothesis of a cataclysm and extinction dealt a blow to any version of the argument from design, the plan of providence, the cornerstone of natural theology or physical theology, which was still a major presence to be reckoned with in the natural history of the eighteenth century. It also contradicted the principle of plenitude, the old Aristotelian idea that was still firmly rooted in the Enlightenment, according to which the world was the best of all possible worlds and all possible forms of existence came into being and did not just disappear like that. Although the earthquake that shook Lisbon on All Saints' Day 1755 had dented any form of rationalizing optimism, planetary extinction went way beyond the case of a localized tragedy, which the fate of Lisbon had eventually been. It even went beyond the grand, if selective and redeeming, punishment that the Flood was taken to be.

The hypothesis of migration to other latitudes enjoyed a larger following. Thanks to the voyages of Captain Cook and those who followed in his wake, many live mollusks were found in the waters of the Pacific and Indian Oceans, of which the

only analogous types known were fossils. These discoveries favored the migration thesis. Founded earlier than and separately
from malacology (the study of mollusks), the study of shells
(conchology) was a fashionable discipline throughout Enlightenment Europe, to which the salons of Paris were no exception,
but it was also a subject on which some of the most interesting
scientific debates of the period were focused.[44] The symmetrical patterns on those petrified carapaces were a wedding
of Art and Nature; invertebrate paleontology produced evolu-
tionist ideas and the notion of biology itself (Lamarck)
(Figure 36). Among the most frequently found fossil specimens
were belemnites, which were not very different from other
known cephalopods. Then there were the nummulites (from
nummus, coin), exquisitely shaped radial and discoidal foramin-
ifera fossils whose accumulation had formed structures like the
cliffs of Dover and whose reticulate interior was the delight of
naturalists and amateurs (a word of noble origin but with pejo-
rative connotations). And of course there were the ammonites,
the familiar Ammon's horns, whose strangeness seemed to
prove the thesis of extinction until nautilus shells began to
arrive in Europe from the East, which seemed rather to support
the thesis of migration and climate change. The same occurred
with the "feather star" crinoid fossils and others from what was
at the time known as the Secondary (later Tertiary) era in Europe. Their counterparts—authentic living fossils—were now
being found in tropical waters. Perhaps the same was true for
the fauna and what had changed was the climate.

The third hypothesis was that of transformism, a speculative
tendency supported more by theory than by any evidence. Its
most extended version during the Revolutionary era was degenerationism. The fact that Juan Bautista Bru had referred to

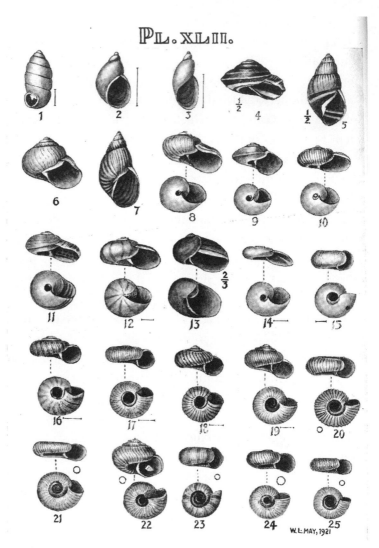

Figure 36. A Lamarckian plate, in W. I. May, *An Illustrated Index of Tasmanian Shells* (1923).

"mutations and changes with the long passing of time" in connection with the Megatherium shows how widespread ideas of this kind were.[45] From the encyclopedist Robinet, via Buffon with his transformism limited to change within the same species, to the painter of the Royal Cabinet of Madrid, there was a wide range of versions and supporters of the theory. They included both philosophers and naturalists, as well as dilettantes without any scientific training worth mentioning. In this sense, it would be wrong to project the victory of evolutionism and natural selection over the transformist theories of the late eighteenth century too much, a common error that all the specialists warn against. There is a big leap from Buffon to Lamarck. Their concepts of what constitutes life are poles apart. And what are we to make of the distance between Bru and Cuvier, or Darwin himself? In terms of training, knowledge, scientific ideas, intellectual stature, and impact, however we look at it, the distance between Bru and the two naturalists is enormous. To say that Cuvier took advantage of the work of Bru is an exaggeration; to regard Bru as a precursor of evolutionism, an act of folly.[46]

In the context of hypotheses regarding fossilized fauna, the big vertebrates soon came to occupy a strategic position for any kind of argument. The chance of discovering new species of cephalopods in the archipelagos of the South Seas (let alone protozoa on the ocean bed) was seriously reduced in the case of large animals. The animal or monster of Maastricht, for instance, provoked great interest. It was discovered in 1766, and its fearsome jaws were exhibited and drawn at the other end of the country in Tylers Museum, Haarlem (Figure 37). It was clearly aquatic, but had it lived in freshwater or the sea? It was initially identified as a gigantic crocodile (the climate of the

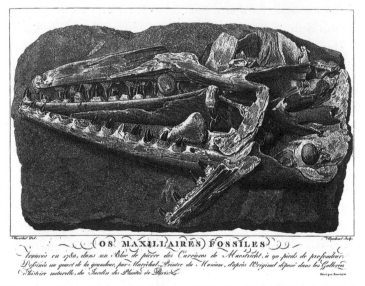

Figure 37. Maastricht Animal jaws, in Faujas de Saint Fond, *Histoire Naturelle de la Montagne de Saint-Pierre de Maestricht* (1799).

remote past was decidedly unrecognizable: tropical reptiles in the swamps of the Low Countries!), until in 1782 the Dutch anatomist Petrus Camper (1722–1789) demonstrated that it resembled more closely a toothed whale, the largest carnivore known, even though it was not exactly a cetacean, a sperm whale, either.[47] In fact, it had characteristics of both types. Was it a whale with the teeth of a crocodile, another impossible animal, a fantastic chimera? Was it extinct, or had it simply escaped detection in some corner of the Pacific?

The question remained unresolved, but it was becoming increasingly clear that fossils held the key to the revolutions of the globe and historical geology. Blumenbach classified fossils as either known or unknown, criteria were gradually established for their dating, and the notion of a history of the earth based on the interpretation of fossil finds emerged. The site in which the animal of Maastricht had been found, a highly solidified limestone riverbed with shells and sea urchins, led to the supposition that it was a relic of marine life dating from before other animals that had been petrified in riverine sediments or in less compact gravel. This was the case of the fossilized teeth, tusks, and bones of the big elephants and rhinoceroses found in various parts of Siberia and Tartary in the eighteenth century. They looked like big pachyderms. Were they? Yes and no. Similar remains discovered in the Alps suggested Hannibal's elephants, but these were so far away, so far to the north. . . . How had they got there?

There was a trade in the petrified ivory of those gigantic tusks in Russia before the age of Catherine the Great (1729–1796). They were collected in St. Petersburg, from where they were transported to the other courts and cities of Europe after being drawn and engraved. The empress invited the German naturalist Peter Simon Pallas (1741–1811) to join the Academy of Sciences in St. Petersburg, where he settled after leading an expedition through the interior of the empire between 1768 and 1774.[48] Pallas collected samples and specimens of plants, living creatures, and fossils from the Urals to Lake Baikal and the Altai mountains. On the banks of a tributary of the Lena in Siberia he found the remains of the skeleton of a rhinoceros, including the skull and even a fragment of its flesh with the skin attached. When he returned to St. Petersburg, he consulted Camper on the similarities between his rhinoceros and those of Camper's own day. He did so according to the common practice at the time: Pallas sent Camper some images to allow him to see it as though he were standing in front of the fossil in question. Pallas believed that the rhinoceros, like the Siberian elephants, had arrived in such northerly latitudes as the result of a planetary flood, the biblical one or some other relatively modern natural catastrophe, a kind of mega tsunami.

As regards the gigantic Siberian elephants, the native people called them "mammoths," the name that would end up being adopted by European naturalists.[49] During the French Revolution many thought that they were antediluvian, and some that they differed from living species. The fossil remains of an extremely woolly skin had even been found on some of them. Were they elephants with a furry hide like a bear or camel? The fact that they were so warmly clad demolished the theory of the gradual cooling down of the planet and the supposed conversion of the northern regions into a tropical zone in the remote past, as Buffon, Daubenton, and others had argued. Or had the climate changed? Had the mammoths developed a natural defense against the cold? That was inconceivable. As one discovery followed another, the theories (and their variants) piled up. The puzzle of the history of life had begun, but it still lacked a unifying logic to put all the pieces and that repository of bones together.

Burtin rather maliciously pointed out the limits of vertebrate paleontology when he adduced the fact that the two best anatomy experts at the time, William Hunter and Petrus Camper, both continued to call the animal from Ohio an *incognitum*.[50] We have already seen the important symbolic role played by this monster from Big Bone Lick, equated with the Siberian mammoth, in the political imagination of the young republic. Buffon considered it to be the only extinct animal; his colleague and protégé Daubenton compared its femur with that of a Siberian elephant and of a living elephant (in a manner of speaking). Were they three different species? Not necessarily, for the difference in size might be due to age, gender, or even diet and way of life. These ideas had a bright future ahead of them, but Cuvier's predecessor, one of the founders of modern comparative anatomy, was not remotely interested in them.

William Hunter (1718–1783), one of the best anatomists of his day, as Burtin claimed, and author of some of the best anatomical illustrations of all time, broke off his research in obstetrics to dedicate himself to a work on the animal from Ohio, which he presented to the Royal Society and published in 1769.[51] He asked his brother John (an important collector and himself a highly qualified physician whose house with a double entrance in Leicester Square inspired Stevenson's *Dr. Jekyll and Mr. Hyde*) for the jaws of an elephant, drew them alongside those of an exemplar from Ohio that had been given to the British Museum, and finally concluded that they belonged to two different species. The American *incognitum* was precisely that, an unknown fossil species that was very probably extinct (Figure 38).[52]

Thomas Jefferson, in his *Notes on the State of Virginia* (1787), argued that the species from Siberia and Ohio were of the same kind: a polar species related to the tropical elephant (the one from Ohio was still classified as a mammoth and had not yet been identified as a mastodon). However, his profound providentialism prevented him from believing that it was extinct. He accepted at face value the native legends about chimerical and gigantic animals (hybrid and magnificent as ever) and supposed that the mastodon-mammoth must still be alive in the unexplored interior of the continent, like the ammonites at the bottom of the ocean. Once he became president, he sent a scientific expedition led by Lewis and Clark (1804–1806) to the

Figure 38. The beast named, successively, Ohio Animal, American *incognitum*, American Mammoth, and Mastodon. Drawing sent by Everard Home to Georges Cuvier, 1804.

Rocky Mountains and the Pacific to find out where two species, as desirable and remote as the exotic fauna was to the Renaissance, were lurking: the symbol of the majestic force of the New World, the American mammoth; and the symbol of its ferocity, the recently discovered "predator," the Megalonyx, that fearsome feline reduced to the status of a gigantic sloth, the North American version of our Megatherium.[53]

Though the stars in these paleontological debates were the big pachyderms, the cases of the Irish elk and the cave bear of Bavaria also aroused interest and raised similar problems: Were they different from their living counterparts or not? Why had they lived in different latitudes and hemispheres? If, as was becoming increasingly clear, many of them belonged to extinct species, what was their relation to present-day ones, and why had they disappeared? It was not just the honor of a nation or the reputation of a continent that was at stake. The different nature of the primitive world and its major catastrophes, the very history of the earth, hinged on them. Vertebrate paleontology was a forensic practice that was unable to reach a conclusion on the identity of those cadavers or on the real causes of their demise. Various solutions were tried out in an attempt to solve the puzzle. Rapid advances were made, although the pieces did not fit entirely together. The discipline was taking its first steps from euphoria arising from discovery to uncertainty about interpretation.

THOUGH IT WAS still too early to resolve these questions completely, what was required at this stage was a sufficiently large empirical base for study and the conditions to allow comparison. These are precisely what Cuvier found when he arrived in Paris in 1795 and obtained his position in the Museum of Natural History. He had at his disposal the largest collection of vertebrates in the world, both alive (in the menagerie next to the Jardin du Roi) and dead (stuffed, or simply in skeletal form). After the centralizing measures and expropriations of the Revolution had augmented the collection,[54] its rooms were populated by skeletons of living species and fossil remains of others that were gradually beginning to be considered extinct, which had been found in deposits that were more recent or closer to the surface than the marine fossils (Primary and Secondary formations in the terminology of the time). As they appeared to be relatively recent, or at least not as ancient as the invertebrates, the possibility of their nonextinction could not be ruled out.

It was not strange for Jefferson to hope that he would find his American wild beasts, or that Viceroy Loreto assembled some *caciques* to ask them if they recognized the mysterious cadaver from the River Luján. North and South America were still to a large extent unknown continents, and very little was known about the sloth, the anteater, and other edentate creatures. True, it was somewhat more than had been known about the Asian rhinoceros in Lisbon in 1515, but it was still not much. As we have seen, the armadillo featured in the iconography of the New World from the sixteenth century on. Another animal with a similar tradition was the sloth, which attracted the attention of the first chroniclers because of its strange, monstrous character. Fernández de Oviedo christened it *perico ligero* ("nimble parrot"—surely an example of Castillian wit). It was described as a small arboreal bear—in fact so arboreal that microscopic algae and lichens clung to its fur as camouflage among the vegetation. This perfect camouflage explains how the slowest mammal in the world was able to survive in the jungle. Depending on the species, it has two or three toes with sharp, hooked claws. This attribute is a recurring feature in the long iconographical tradition that extends from the descriptions of America by Thevet or De Bry, the teratological work of Paré and the natural histories of Gessner and L'Écluse to Nieremberg and Jonston in the seventeenth century. It even featured in Kircher's *Musurgia* for its song, which was said to reproduce the heptatonic musical scale![55]

In the course of the eighteenth century, however, the sloth began to lose its glamour. Buffon had passed sentence on it: its torpidity and exasperating pace had turned it into an example of the degradation of American nature (Figure 39).[56] There was a skeleton of this species in the Museum of Natural History in Paris, making it possibly the only place in Europe where Cuvier would have been able to identify our exemplar.

Plate XXXVIII.   *Page* 124 Vol. II.

Racoon.

Two Toed Sloth.

Black Coati.

Great Ant-eater.

Figure 39. Georges-Louis Leclerc, Comte de Buffon, *Histoire Naturelle, générale et particulière* (1749–1788).

At the sight of the set of plates that arrived from Madrid and their comparison with the skeletons available in the museum, it cannot have taken him long to solve the riddle. An herbivore with rear limbs that were longer than the front ones, with similar clavicles, and with hands and feet ending in three toes with hooked claws? What other skull had that characteristic prominent lower jaw and such a large apophysis at the base of the zygomatic arch? Though on a different scale and with certain differences, the skeleton in the Madrid Cabinet was clearly related to the sloth and other edentates. In his seminal work on the Megatherium, Cuvier combined his anatomical knowledge with the power of images to demonstrate it. An illustration showed the skulls of the Megatherium, a two-toed sloth, and a three-toed sloth. The similarities were striking (see Figure 29).[57]

Cuvier also worked with great circumspection. He knew who he wanted to become, but he also knew who he still was, so he took great pains to pay tribute to the founders of comparative osteology (Hunter, Camper, and especially Daubenton, who was still alive). The respective fields of expertise were parceled out in Paris at the time: geology was the domain of Barthélemy Faujas de Saint-Fond, invertebrates that of Jean-Baptiste Lamarck, reptiles that of Bernard Germain de Lacépède, mammals and birds that of Geoffroy Saint-Hilaire. Cuvier soon joined with the latter to study pachyderms: the two-horned rhinoceros and the elephants. Cuvier reclassified them, including tapirs and pigs in the genus. During the months in which he was working on the Megatherium he also commenced another pioneering study of living and fossil elephants. Our two protagonists, the rhinoceros and the Megatherium, always appear in the company of or preceded by the elephant, as if to give them (or us) the scope and scale of their possibilities and forms.

As he had done with the Megatherium, Cuvier first delivered a lecture on living and fossil elephants at the National Institute, before publishing it a few months later in the *Magasin Encyclopédique*.[58] He boldly asserted that there were not three but four species of elephant: two fossil species (Siberia and Ohio) and two living species (Asian and African). Detailed anatomical study of their jawbones demonstrated their important differences. They belonged to the same genus, but were distinct species. One was not derived from the other, there were no predecessors or successors. The living species belonged to the present world and had emerged (or been created) after a planetary cataclysm; the fossil species were not just traces or testimonies; they were the petrified witnesses of a remote world that had disappeared for ever. Cuvier was gradually adopting the catastrophism and binary model of Deluc (the thesis of two distinct worlds), affected too by Blumenbach's concept of the *Total revolution*. Part of his bad reputation in Whig historiography is due to his defense of this thesis, which Lyell, Lamarck, Alfred Russel Wallace, and Darwin, with the hindsight of uniformism and natural selection, were to disprove (he might have been treated more kindly if the generations immediately after Darwin had known of ice ages, the impact of giant meteorites, and the theory of "punctuated equilibria").

But it is not the historian's task to celebrate the ideas that triumph or to deprecate those that fail to prosper. Neither should we extend anticipated baptisms or funerals. Cuvier burst upon the scene of vertebrate paleontology with those pioneering works and became a major figure in the discipline after the publication of the *Recherches sur les ossemens fossiles de quadrupèdes* (1812), the magnificent work in which he brought together his paleontological research. When he died in 1832, it was a mere six months since the *Beagle* had set sail from Plymouth.

In fact, Cuvier's science was marked by the geometrical and mathematical model of Laplace. He understood animals as genuine machines whose parts were connected and interdependent. In his early works on the Megatherium and the elephants, he began to apply the two principles that—to paraphrase Newton—might be called mathematical principles of zoological osteology. The first of these, "conditions of existence," "final causes," or "correlation of parts," justifies the presence of a part or an organ and its relations with the rest in a functional, teleological and finalist way: living creatures are composed of a series of parts and interrelated organs in such a way that the composition permits the existence of that animal in a given habitat. In accordance with the old Aristotelian motto, nature did not do anything in vain. A carnivore needed incisors to shred the meat, a stomach to digest it, jaws or claws, and the ability to move quickly and to hunt its prey. The other principle, the subordination of characters, followed from the first: the parts were arranged hierarchically to indicate that some were more important for existence and that they determined the others. Given the dependence or subordination between them, once some necessary conditions were known, the rest could be deduced. As the reader can imagine, this was of great importance in a practice like vertebrate paleontology. "Give me a tooth and I will reconstruct a dinosaur": the fantasy of every paleontologist and the one popularized in a well-known film using a fragment of DNA.[59]

The five volumes of the *Leçons d'anatomie comparée* were published between 1800 and 1805. This work refounded a discipline that went back to Aristotle and had been cultivated during the Renaissance by physicians like Pierre Belon and during the Baroque by others such as Claude Perrault, who had already postulated a Cartesian-style animal mechanics. Cuvier gave this tradition a new orientation and catapulted it toward the work of his opponent, Geoffroy Saint-Hilaire, and the works of Richard Owen. Nor should we forget Ernst Haeckel, the German biologist who assumed the task of incorporating anatomy in the theory of evolution and who also produced one of the most original works on the relation between science and art, the lithographs in his *Kunstformen der Natur* (1904).[60] What interests us here, however, is how Cuvier used comparative anatomy for the geology of vertebrate paleontology, a specific application undertaken to demonstrate that the fossils discovered belonged to extinct animals.

Soon after his work on the Megatherium, Cuvier had other triumphs, aided in part by Napoleon's plundering of European natural science collections to augment the holdings of the Museum of Natural History (it is not only works of art that are seized in wartime). Valuable specimens arrived from present-day Belgium and the Netherlands, such as the skulls of elephants from Ceylon and the Cape of Good Hope represented in his 1799 article. Cuvier did important work on the Siberian mammoth and the animal from Ohio—the American mastodon, as he ended up naming it in 1806.[61] He showed that the Bavarian cave bears did not resemble living polar bears. And the elk fossilized in the peat bogs of Ireland (*Megaloceros*) was not of the same species as the living Scandinavian elk. Cuvier discovered that there were extinct species of hippopotamuses (in fact, far more than the two species of living ones). He also distinguished a further four species between the living and fossil rhinoceroses (they could not be fewer in number than their time-hallowed rivals, the elephants).[62] In 1798 he had the opportunity to enroll in Napoleon's Egyptian campaign but preferred to stay where he was. It was his mentor and colleague Saint-Hilaire who took part in one of the most famous military expeditions of all time. Why did Cuvier decide to remain in Paris?

There were several reasons. Why should he move from that privileged place that was at the receiving end of all the pieces, fossils, bones, drawings, plates, news, reports, and publications that one could desire? Cuvier was fortunate to be at the center, where everything came together. There was no need for him to go and see the world; the world came to see him. He could see more fossil bones and have a better scientific career—his two prime concerns—in Paris than anywhere else.

It could also be said that Cuvier preferred the low-profile isolation of the hermit to the more gregarious work of acquiring firsthand empirical and sensorial knowledge of the world. If it came to choosing between these two archetypal poles in the acquisition of scientific knowledge, between fieldwork and work in the cabinet, or as we would say today, the laboratory, Cuvier opted for the remoteness of the recluse.[63] In a certain way he occupies the polar opposite of knowledge *ad vivum*, of the ideal of coming into contact with artifacts in the field, the strategy of travelers as readers of the Book of Nature and which Dürer adopted (or pretended to adopt, as in the case of the rhinoceros) in his representations of the natural world. Real knowledge does not come from touching and feeling things in some wild (and in the case of fossils, anachronistic) context, but on the contrary, from not being in situ but able to abstract, isolate, compare, and reconstruct—to imagine instead of to see, though Cuvier would never have openly admitted it.

Here lies another paradox of his method and his fortune, because the most geometrical of the naturalists was also the most scrupulous of observers. While his science had the style of the mathematical axiom, his style was outspokenly empiricist. His work is the exemplary culmination of the two major traditions of the scientific revolution—mathematics and experiment—applied to the field of anatomy. His staunch defense of the binary model (the division of natural history into two periods representing different worlds) and the stability of species, the refusal to countenance any version of transformism or degenerationism, had an empirical foundation. There was no evidence, there were no proofs in their favor. The facts themselves spoke of living and extinct animals, of the profound and irredeemable differentness of the lost world. All the rest was conjecture and speculation, which Cuvier avoided like the plague. Did the laws of geology continue to operate in the present as they had in the past, as the uniformists believed? It was impossible to know, Cuvier replied, sticking to the facts like an entomologist, collector, or antiquarian. On one occasion he stated that he left grand geological theories to the "great geniuses," thereby expressing his disdain for the indemonstrable hypotheses that had accompanied sacred physics and theories of the earth.[64] Cuvier was the last great Aristotelian naturalist in that his universe was taxonomical and logical; his science paid tribute to what was singular and specific, to what was observable. Species could not disappear or be confused: there was no evidence. What was needed to make out the action of ordinary physical processes or modification in the living species in the course of vast quantities of time was not more facts and more *ad vivum* observations, but the opposite: more distance, more inferences, more imagination, that ensemble of mental faculties that are necessary to approach the scale of deep time in any way, to understand the relations of living creatures with their habitat, the effects of domestication or the mechanisms of natural selection. None of these is a visible phenomenon, or one that can be easily observed.

It was not only from remote regions that fossils arrived in Paris, the metropolis of 1800. The capital's surroundings too held their surprises, such as the bones of strange creatures encrusted in the sedimented gypsum quarries of Montmartre. Cuvier reconstructed one of them from scratch. In other words, he had to imagine it and give it a shape before identifying it as a new unknown species. In this it resembled the Megatherium, although in this case there were no prior drawings from elsewhere, no previously attempted assembly. Once he managed to draw it or imagine it—he would have said deduce it—it appeared as an ungulate with a short trunk. It had something of the camel and quite a lot of the tapir, a hybrid between the pachyderms and certain ruminants. He called it a Paleotherium, emphasizing its antiquity, but also recalling the object of his activity: the past.

Cuvier regarded himself as an antiquarian of extinct fauna, "antiquaire d'une espèce nouvelle" as he called himself in the Preliminary Discourse to *Recherches sur les ossemens fossiles* (1812), a masterpiece in the context of his intellectual production and of the grand scientific rhetoric of the period that was clearly modelled on the Preliminary Discourse par excellence, that of D'Alembert in the *Encyclopédie*. His task, as he saw it, was to restore those monuments of past revolutions and to decipher their meaning.[65] If the Alps or volcanoes were the monumental ruins of a relatively recent catastrophe (as Deluc, Saussure, and Déodat de Dolomieu believed), and if ammonites and nummulites were the coins or medals of an extinct fauna, or at least an unknown one, as the conchologists noted, why not extend the same analogies to the large vertebrate fossils? Paleontology was emerging at the crossroads of geology and biology, two disciplines that were coming into their own at the same time. However, it is important to insist on what Martin Rudwick has demonstrated: Cuvier understood it as a practice akin to antiquarianism, collecting, or history.[66] It was a question of reconstructing skeletons from a few remains, like the fragments of vases or temple friezes. He was a historian of fauna, but a particular type of historian, one who studied the very ancient (prehistoric) past that was nevertheless comparatively recent (the notion of an antediluvian, pre-Adamite world had been formed only a few decades earlier) on the basis of material objects, not words or written sources. He was an archaeologist of forms of nonhuman life that had been lost and finally recovered.

His science was thus a kind of zoological archaeology that followed an empirical method and had mathematical pretensions. Its predictive character was staged in those same quarries of Montmartre when the bones of another remarkable cadaver were exhumed in various stages. First came the head. He deduced from its dentition that it would have the classic marsupial bones in the lower part of the column at the level of the pelvis (Figure 40). It did. When the hips of the fossil were extracted, Cuvier's reaction was as if Halley's comet had crossed the sky on the day predicted. Against Kant, who denied the life sciences the status of celestial mechanics, there was no science, according to Cuvier, that could not become almost geometrical.[67]

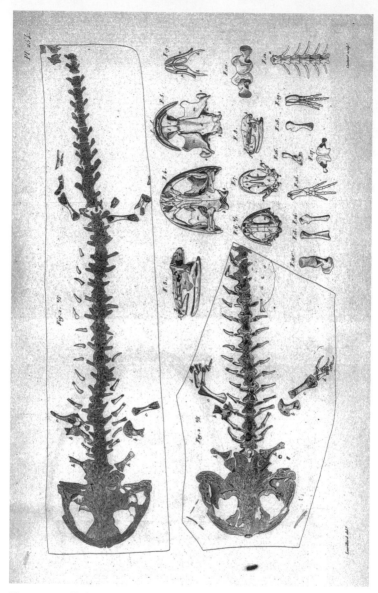

Figure 40. Skeleton of a terrestrial salamander. Its reconstruction and that of the Montmartre marsupial caused similar problems. Georges Cuvier, *Recherches sur les ossemens fossiles*, 4th ed., Atlas (1836), Vol. 2, Plate 254.

In this archaeological and deductive task, a full-fledged forensic anatomy did not end once it had identified, named, and illustrated the deceased. In the last instance, the anatomy of a living being—its form, its image—revealed its habits and customs, its diet, and the life it had led. Within Cuvier's functionalist and finalist perspective, everything—every bone, every piece, every organ—had a purpose. Once it had been established what it was, its use could be deduced from that. Life could be inferred from anatomy, or—to invert the history of our rhinoceros—once the image had been created, the words, discourse, and predicates could be accumulated and elaborated as Plinian natural history was brought to bear on the Megatherium and all those other strange inhabitants of a lost world. Whereas legend and word dominated and preceded image in the case of the rhinoceros, it was the images of the Megatherium and other extinct animals that generated words, histories, narratives.

Cuvier's work, like that of every paleontologist and every historian, bore within it a desire for regeneration, almost for the resurrection of flesh or life. Once they had been found, those fossilized, fragmentary, and dispersed bones came to be drawn and represented rather like a zoological *vanitas*. It is difficult to ignore our culturally embedded symbolic reactions when confronted with a skeleton or, even more, a skull.[68] More than any other part of the anatomy, the skull is where life does or does not reside, in the gaze, the most obvious indicator of death. Although, or perhaps because, they are the bones or skulls of animals, they are all the more disturbing because they evoke a basic resemblance between living beings, whether we consider their skeletal structure or their—and our—final destiny (Figure 41).

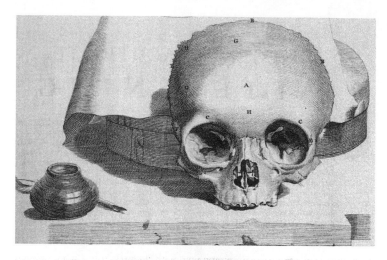

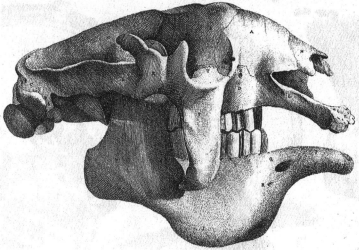

Figure 41. *Vanitas*. Human skull *(above)*, in Govert Bidloo, *Anatomia Humani Corporis* (1685), and skull of the Megatherium *(below)*, by Bru and Navarro, in José Garriga, *Descripción del esqueleto de un cuadrúpedo . . .* (1796).

Seen in this light, vertebrate paleontology can be understood as an activity connected with the regeneration or restitution of skin and flesh. It was thus like taxidermy, that technique we saw linked with taxonomy in connection with Bru's activities in the Royal Cabinet in Madrid. But there is a regeneration of a different kind too: the reconstruction of an animal's habits and customs. The restoration of a creature's life is not complete without the recovery of its moral history as the humanists understood the term, the recovery of their social history or ethology, the totality of their habits and forms of life. This is no metaphor, as Cuvier made abundantly clear:

> The bones being well known, it would not be impossible to determine the form of the muscles that were attached to them; for these forms necessarily depend on those of the bones and their ridges. The flesh being once reconstructed, it would be straightforward to draw them covered by skin, and one would thus have an image not only of the skeleton that still exists but of the entire animal as it existed in the past. One could even, with a little more boldness, guess some of its habits; for the habits of any kind of animal depend on its organisation, and if one knows the former one can deduce the latter.[69]

It is within this context of regeneration of life, not limited to biology, taxonomy, or anatomy but including the social, environmental, and historical lives of animals, that we have to place the role of anatomical drawings and illustrations. They serve a double purpose: they form part of the scientific argumentation as an instrument of visual rhetoric to demonstrate certain ideas and to persuade both colleagues and the public; but they are also a strategy to animate life, to restore what those monsters of the lost world no longer have, thereby closing the circle of a virtual, fictitious, imagined resurrection.

The image of the Megatherium in the successive editions of the *Recherches* became more dynamic and lifelike as it left behind the static character of Bru's images (Figure 42). This required the services of another draftsman and naturalist who saw it and represented it *ad vivum*. As an extraordinary pioneer in the representation of prehistoric fauna, Cuvier perfected his technique over the years, and as he worked on other exemplars and cases he developed an iconographic animation program that was to be continued by all those who came after him.[70] Combining continuous and dotted lines to indicate which bones were real and which were hypothetical, which had been found and which had been deduced, Cuvier managed to

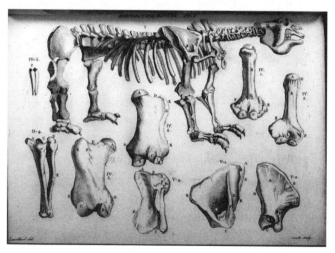

Figure 42. Georges Cuvier, *Megatherium fossile*, in *Recherches sur les ossemens fossiles* . . . (1812 ed.), vol. 4.

illustrate some species in motion with a suspended or bent limb as George Stubbs had done in the case of the horse. There is no life without motion. If the quality of a cadaver is stability, tranquility, as in a *vanitas* or still life, the quality of living beings is the opposite. If fossils are the result of the petrification of organisms over time, the desire both of paleontologists and of historians is to animate them (Figure 43).

In fact, like Dürer, Cuvier made a living from illustrations, both from making them and from receiving them from different parts of the globe. The extremely sedentary naturalist established an extensive network of correspondents, thanks to which he received drawings of dental pieces, jawbones, and femurs from Siberia to Ohio and from South Africa to Scandinavia. Camper's son sent him the drawings that his father had made of the animal from Maastricht; Giovanni Fabbroni sent him some molars from the Florentine museum that looked like those of a hippopotamus; a British anatomist sent him the image of the American "mammoth" that Charles Wilson Peale was exhibiting in London in 1802, which enabled Cuvier to reveal the mystery of the *Incognitum Americanum* and to correct the persistent error of identification by calling it by the name that is still in use today: mastodon.[71]

The noblest commerce of which a human being could boast—*commerce* being an enlightened metaphor that combined sociability, exchange, and knowledge—consisted of mobilizing objects at a distance in three possible forms: through traffic in objects themselves (fossils traveled relatively easily, and even large ones like the mastodon could be dismantled and transported); via the exchange of letters (an extension of conversation, Foucault's "the background noise, the endless murmur of nature," Hamlet's "words, words, words"); and by

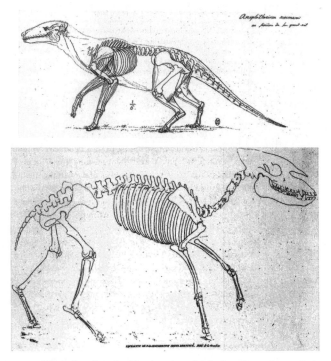

Figure 43. Two sketches by Cuvier, in which he shows his gift for the animated restoration of the big extinct vertebrates. *Anoplotherium Commune (above)* and *Palaeotherium Minus (below)*. The former is in the Museum of Natural History (Paris) and the latter is from a copy in the University Library (Cambridge).

the spread of images of fossils, that is, their copies.[72] The images were—and still are—the most useful and the most commonly used of the three in scientific exchange and the transfer of knowledge. Without having to study zoological anatomy or paleontology, all you have to do is switch on a television or connect to the Internet to see what is the most effective, immediate, and lifelike way to transfer a thing from one extreme of the globe to the other.

From the very first, paleontology was a visual and collaborative science, as it had been in the days of Gessner and Steno, but it was Cuvier who formalized both aspects and gave them a modern, institutional, regulated, almost industrial impulse. He used the *Magasin Encyclopédie*, the *Journal de physique, de l'histoire naturelle, et des arts*, and other journals to issue calls for international cooperation in a collective effort, though he ended up controlling it all himself (every army has its generals). The Museum of Natural History became a factory of chimerical and fascinating images of extinct big vertebrates, of which Cuvier managed to recover more than twenty. The unknown monsters conquered the public imagination. That noble commerce between the distinguished members of the Republic of Letters was taking on new roles, like the ones that Peale was exploring and beginning to exploit in Philadelphia and in his European tour with the mastodon, a traveling circus that recalls inescapably the grand tour of the Asian rhinoceros (1741–1758) and the peregrinations of Ganda.

As we have seen, the exhibition of extraordinary fauna had a long tradition. What was new this time was their origin in the remote past and the social dimension of the phenomenon. The display of remains of the lost world, the most exotic fauna imaginable, was turning into a business venture and a public spectacle. The history of life on this planet emerged at the time when new civic ideals were taking root. Access for every individual to the world of science was an implicit and characteristic promise of the Enlightenment. The Republic of Letters could expand indefinitely (though nobody could have foreseen the places and forms in which it is continued today). The earliest forms of mass culture were around the corner, to be witnessed by the Megatherium from its pedestal in the Royal Cabinet. When Queen Victoria inaugurated the Great Exhibition in the Crystal Palace in London in 1851, models of three reconstructed dinosaurs in lifelike poses decorated the grounds (Figure 44).

It is impossible to do full justice to Cuvier here. His defense of the creation, catastrophism, and stability of the species have inevitably tarnished his reputation. At the time, the alternative to his ideas about fossils was the transformism of Jean-Baptiste Lamarck, of which the same could be said of Cuvier.[73] Triumphalist historiography usually focuses on Lamarck's ridiculed idea of evolution based on the inheritance of acquired characteristics (a hypothesis that could only occur to a brilliant thinker!). What is certain is that the great creator of invertebrate paleontology was able to read in fossil and living shells the slight, constant, and almost imperceptible changes in organisms over the course of time and under the influence of the environment (see Figure 36). Although still immersed in the speculative language and vitalism of Enlightenment natural philosophy, Lamarck understood, as Lyell, Wallace, and Darwin would understand a little later, that time meant almost nothing to nature.

Cuvier energetically defended his thesis. He deployed his entire arsenal of arguments, his vast menagerie of fossils, and his immense talent to demonstrate that species were stable. One of the finest moments in this intellectual battle is his study of the mummified Egyptian ibis.[74] Never was this naturalist more of a historian: he drew on Herodotus, contradicted Pliny, and consulted Hebrew and classical sources to show that the bird mummified more than three thousand years earlier was identical to the *Numenius Ibis*, a living species commonly known as

Figure 44. The studio of Benjamin Waterhouse Hawkins in Sydenham, where he made the Crystal Palace dinosaurs.

the sacred ibis. There were no signs of transmutation, at least not during the presence of human life on earth. Cuvier could not know that three thousand years were a trifling amount in the history of life. The Napoleonic campaigns and the funerary rites of ancient Egypt were put to the service of a comparative anatomy that in turn served paleontological and geological ends.[75] What is more fascinating than the history of how we came to know what we know?

WHAT HAPPENED TO our American monster? The Megatherium unleashed a stampede of antediluvian fauna in the imagination of the West, which is sufficient reason for it to occupy a distinguished place in a hypothetical museum of our representations of the earth's past. But its biography, or rather its many lives, do not stop here. This second part of our essay cannot conclude with Cuvier's identification, with his naming and drawing of it as an edentate that was closer to a sloth than to an anteater in 1796. Nor can it even end eight years later when he wrote an extended version of that pioneering work and published it in the *Annales du Muséum d'Histoire Naturelle* (1804), pages that he later incorporated in his *Recherches sur les ossemens fossiles* (1812).[76]

Megatherium fossile: determining a species, classifying it in its niche, was like stuffing or petrifying it, placing it in its coffin. The property of a skeleton is stability. This one had been stable, at least in its petrified form, for thousands of years, but now, recently resuscitated, it seemed to be making up for lost time and living not one but several lives or metamorphoses.[77] This fossil vertebrate was endowed with a tremendous flexibility. Its versatility marks it as a truly modern subject that resists classification, tabulation, being confined to a fixed place and a single identity.[78]

Cuvier summed it up in the title of his 1804 article in the *Annales*. Its title contains a reference to our own essay: "On the Megatherium, another animal of the sloth family, but of the size of a rhinoceros. . . ." It included illustrations and a translation of Garriga's edition done by Aimé Bonpland, the French botanist and Alexander von Humboldt's traveling companion in his voyage to South America. Cuvier deduced the animal's habits from the most determinant characteristics of this or of any other species: its dentition and its toes. Its teeth showed that it was an herbivore; its robust front paws, armed with sharp claws, would have been used to attack roots and branches, as well as to defend itself. "It was not a fast runner" deduced Cuvier, always ready to rationalize any presence or absences, because, "it did not need either to pursue or to escape."[79]

We have already seen how, while Cuvier was christening the animal in 1796, Jefferson almost converted the Magalonyx—its northern cousin—into a terrible feline, a carnivorous predator whose ferocity, he hoped, would symbolize that of the New World. In 1804, when Cuvier was completing his first resurrection, President Jefferson was dispatching the Lewis and Clark expedition to the Rocky Mountains to look for survivors among the *americana incognita*. Faujas de Saint-Fond (1741–1819), the successor of Buffon and opponent of Cuvier on geological matters, regarded the Megatherium too as a huge carnivore. How could it be compared with the "unfortunate and indolent sloth"?[80] At the same time, in Madrid the Megatherium was playing a role in the libretto of the debate on Spanish science.

It could be adapted to a patriotic discourse just as easily as it could defend the nature of a continent, the emergent republic, or the wounded honor of a glorious and slighted monarchy.

Besides Philippe Rose Roume and William Carmichael, whom we encountered in Chapter 5 in the role of the "Moravian trader" of our first story, a third person visited Madrid and saw the Megatherium before the century was out. Peter Christian Albidgaard, a Danish astronomer and collector, studied and drew the skeleton in the Royal Cabinet in 1793. He did not send this information to anyone, but used it to publish a small report in his native Danish four years later that included an illustration of the skull and some of its individual bones. Albidgaard assigned it a place among the *bruta* of Linnaeus, an order that Cuvier had modified to create the edentates. It is one more small episode in this first phase of the interpretation of the Megatherium covering roughly three decades (from the 1790s to the 1820s). (Historians and paleontologists always end up measuring time through periods instead of dates: the limit of the Cretaceous, beyond the late Empire, and so forth.) Three moments could be singled out in this early phase: the discovery and probable first reconstruction of the unidentified beast in Río de la Plata; the assembly and the plates of Juan Bautista Bru and Manuel Navarro in Madrid; and Cuvier's work in Paris, which put an end to the speculations and doubts and undeniably exercised its sway down to the 1820s (and in a certain sense continues to do so today). Historians and naturalists both like taxonomies (of time and of life respectively). It was neither a giant nor a big carnivorous feline lurking in the interior of the pampas. It was a gigantic and extinct edentate, an intermediary species between a sloth, an anteater, and an armadillo. Its second metamorphosis is connected with its capacity to move even within that ontological boundary, in the morphological diversity that the boundary allows, which is greater than might appear at first sight.

Interest in the Megatherium revived once more at the beginning of the 1820s with the publication of the first volume of a work on comparative osteology, *Die Vergleichende Osteologie der Säugetiere*, the result of collaboration between the Russian embryologist and epigenist Heinz Christian Pander (1794–1865) and the German naturalist and engraver Joseph W. Edward d'Alton (1772–1840).[81] In 1818 the two scientists had decided to visit a number of European museums in search of proofs and arguments for their transformist ideas. They visited Madrid and were so fascinated by the Megatherium in the Royal Cabinet that they devoted the first volume of their work to "the giant sloth, described and compared with other related species." Inevitably, their anatomical description of the big vertebrate followed Cuvier, but they interpreted it from the viewpoint of the *Naturphilosophie* that was characteristic of the period in Germany, some of whose concerns have been flitting around this essay from the first page: the hidden morphological relations and correspondences between natural objects.

Pander and d'Alton considered the Megatherium and the Megalonyx not as predecessors of the slow-moving edentates but as their ancestors. In a wave of German inspiration (the only kind able to rehabilitate the *physis* of the Greeks), they transposed Goethe's conception of metamorphosis from the world of plants to the whole body of living beings as well as to its individual members. Caspar Friedrich Wolff had referred to the *vis essentialis*, and more recently Blumenbach had spoken of *Bildungstrieb*. On the basis of his vitalist epigenism, Pander explained the mutations in the development of an individual and the changes within a species as a whole by resorting to the notion of metamorphosis that had already been employed by Petrus Camper, another champion of comparative anatomy. This version, however, was explicitly based on Goethe and his ideas about the archetypal forms that are manifested and evolve in a constant formative process *(Bildung)*.[82]

So now the Megatherium was the ancestor of today's sloth, into which it had been transformed. It had diminished in size and taken to the trees, very probably because of a generalized rise in the water level, a flood stabilized in a massive and permanent inundation. Was this such a strange idea? It depends on how it is treated; every intelligent reader can imagine its various versions. Anyone who is well informed knows that the first forms of life were aquatic, and that extinction and adaptation, evolution over billions of years and some (occasional and major) natural crises have not been precisely incompatible phenomena. Pander was one of the founders of modern embryology and one of the first to make an association as brilliant as it was fertile: the association between the paleontological record, phylogeny, and ontogenic development. *Pace* Cuvier, the species were not stable. That is certainly true: nature is not a given reality, allocated its niche, whether coffin or pedestal, but something in motion, a *natura naturans*. And history cannot be understood as a lapidary or monumental discipline; the past is not something that is static and over.

In fact, Pander and d'Alton had read *Versuch die Metamorphose der Pflanzen zu erklären* (1790) by Johann Wolfgang von Goethe (1749–1832), that anti-Linnaean change of course forged in the gardens of Italy in an attempt to unravel different laws of life and different forms of knowledge.[83] If his ideas on the archetype and metamorphosis influenced Pander and d'Alton,

Goethe could not help agreeing with what he read in their *Die vergleichende Osteologie,* the first two volumes of which he reviewed and highly praised. The genius of German letters, inexhaustible polymath, and man of science capable of refuting Newton himself managed to use the Megatherium to outline an idealist evolutionary history. Our antediluvian creature, he thought, must have descended from a marine mammal that had been obliged to live in the swamps. There, Goethe argued, it would have developed limbs to enable it to move over the marshy earth. This theory was not to be taken literally. It was both a poetic and a historical way of reconstructing nature, since it tackled nature from a conception that depended less on fixed forms *(Gestalten)* than on the process of formation *(Bildung)* and becoming. These ideas were shared by Herder earlier, by Schelling later, and were in a way legitimized by Darwin himself, who mentioned Goethe as one of his predecessors, though whether from genuine admiration or in an attempt to legitimize or ennoble his own ideas remains unclear; to invoke Goethe was like appealing to the new Homer.[84]

The Megatherium of Pander and d'Alton also underwent a slight mutation. The suitability of its feet and claws for digging revealed the existence of an animal that did not merely root around in the soil occasionally, but was an expert constructor of trenches, a creature with burrowing habits. Although its clavicles seemed to suggest an arboreal creature, its toes showed that it could not have lived suspended from the trees.[85] He moved a slight distance away from the sloth *(Bradypus)* and came closer to the armadillo *(Dasypus).* Like the rhinoceros of Damião de Góis, the Megatherium seemed to be associated with the earth, but with its depths, not its surface. In a certain sense, the fossil returned to where it had been found. According

Figure 45. H. C. Pander and J. d'Alton, *Das Riesen-Faulthier, Bradypus giganteus* (1821).

to this version of German transformism, the descendant of a magnificent whale was turning into . . . a gigantic mole (Figure 45)!

When Cuvier prepared a new edition of the *Recherches* in 1823, he had to slightly modify his conception of the big edentate. He replaced the images of Bru and Navarro with those of Pander and d'Alton, which thereby became independent of the text and the embryological theses in which they had been embedded.[86] These images, like any others, have a life of their own, some very long and expansive (like Dürer's rhinoceros), others less so. The plates of Bru and Navarro were beginning to appear deceitful, buried beneath more recent ones that were more dynamic and more teratological (Figure 46). The Mega-

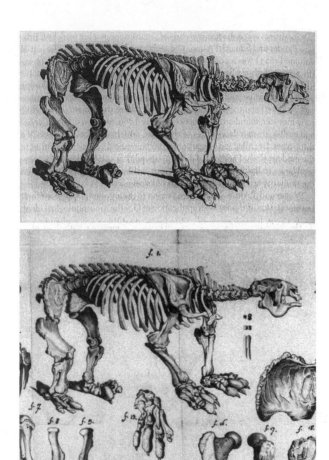

Figure 46. The Megatherium, in H. C. Pander and J. d'Alton, *Das Riesen-Faulthier* (1821) (*above*), and Georges Cuvier, *Recherches sur les ossemens fossiles . . .* (1836 ed.) (*below*).

therium was no carnivore, but it was beginning to flex its forelegs. D'Alton enjoyed a great reputation as an illustrator of images of nature at the time. His magnificent two-volume monograph on the horse had placed him in the wake of Stubbs.[87] As a professor of natural history, painter, and consummate engraver, d'Alton produced some highly evocative images of the Megatherium. By animating the skeleton, he made it look more rather than less fierce. He was very interested in representing motion, as well as other processes or phenomena that were difficult to capture, invisible, or otherwise obscure. For instance, he drew the stages of the embryological development of the chicken, a work of great artistic and scientific importance that drew on a classical motif within the arguments of Karl Ernst von Baer and Pander himself, the founders of embryology. The illustration of extinct fauna was the ideal field for him to develop further his talents and interests. It provided another opportunity for him to give form to natural phenomena that could not be observed conventionally, if at all. The representation of prehistoric epochs was to become a genre in its own right in the following decades.[88] After the irruption of that vast amount of deep time, there was a need to draw scenes of the alien past, the pre-Adamite world, the extinct forms of life, everything that had never been seen or represented by human hand or eye. It was necessary to imagine the invisible.

Thanks to the work of Pander and d'Alton, Cuvier was able to confirm that Bru had been mistaken about the distribution of the toes of the feet (or front paws). Only three of them had hooked claws. The thumb did not have one, and the little finger was hidden beneath the skin (Figure 47). In the new edition of the *Recherches*, Cuvier even gave credit to Saint-Hilaire, his great rival in the field of comparative anatomy, when he pointed in the same direction as Pander and d'Alton: perhaps the Megatherium was more like an armadillo than a sloth? Did it have armor?

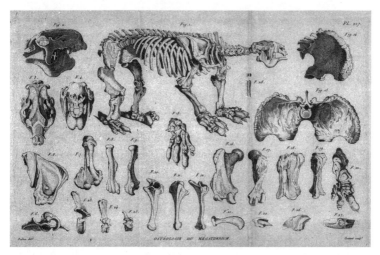

Figure 47. Georges Cuvier, *Recherches sur les ossemens fossiles . . .* (1836 ed.), Atlas, vol. 2, plate 217.

It was a disturbing question. Cuvier skated around it, but some of his followers raised it, which leads us to the finale of our story. The possibility that the Megatherium was protected by a kind of chain mail is the final act in this double circular history. Did it have armor? Besides sharp claws, fantastic beasts need defense. Never did our Pleistocene edentate come so close to the rhinoceros. The desperate search for what it lacked (skin, legend, habits, history, life) accorded it a brief existence with the plated hide of the big armored creatures.

The Megatherium had burst onto the scene with its first resurrection in Madrid, and now it starred in the debate between Cuvier and Geoffroy Saint-Hilaire. This dispute marks the apex of comparative anatomy in the 1820s and 1830s and was continued after the death of Cuvier in 1832 by the defenders of his functionalist naturalism (based on anatomical resemblances and analogies between related groups and on finalist explanations) and the followers of the structuralism of Saint-Hilaire (based on the structural analogies of all living things, and in which anatomy was taken to determine habits rather than vice versa).[89] Does function create the organ or the other way around? Does every feature serve some purpose? Why are certain parts of organisms atrophied or hypertrophied? What was the similarity between cephalopods and vertebrates? The questions did not cease.

The Megatherium, a creature with a strange anatomy, "the king of the monsters" as another of its students, Woodbine Parish, called it, had been a key piece in the zoological osteology of Cuvier. Through a strange inversion of the arrow of time, it had inherited the characteristics of the sloth. However, gradually, after its descent from the trees and its closer approximation to the anteater or the armadillo, other anatomical features required an explanation. What were they for? And if they served no useful purpose, did not that contradict the finalist anatomy of the scientist who had given it its name? The Megatherium could turn into the Trojan horse of the model that it had helped to create. Its malleability made it difficult to domesticate and classify. It did not fit in with the argument from design or natural theology at all. It had no place in either the plans of providence or the functionalism of Cuvier. Monsters, the ancient markers of the heavens, living proofs of god, were now able to reveal the inefficacy and lack of intelligence of the Creator, perhaps even calling into doubt his very existence.

In 1823 the remains of armor belonging to an animal related to the Megatherium arrived in Paris. It seemed to be a gigantic armadillo *(Dasypus)*.[90] Other remains of a similar kind had been found at different points of the Río de la Plata. *Bradypus* or *Dasypus?* Did all these remains belong to one species or two species? Was the Megatherium a *Bradypus* (a sloth) or a *Dasypus* (an armadillo), or a hybrid of both? Might the Megatherium have had armor, like an armadillo? Fantastic zoology continued to produce its chimeras as it deployed every possibility of the art of combination and hybridization. Just as the rhinoceros had been the sum total of the body of a bull, the head of a wild boar, the feet of an elephant, and the horn of a unicorn, our creature now became something with the size of a pachyderm, the head and teeth of a sloth, the incipient trunk of an anteater, or perhaps of a tapir, and the claws and armor of an armadillo.

Rev. William Buckland (1784–1856) gave the penultimate twist to this story of the big extinct and transformist vertebrate. The first holder of the chair of geology at the University of Oxford was a brilliant and decidedly British personality (if Kircher seems to have stepped out of a story by Borges, Buckland is worthy of the Pickwick Club). An extreme empiricist, Buckland used his five senses to the full to examine natural phenomena: he went so far as to taste the urine of a bat and even seemed to have nibbled a royal French relic, the heart of Louis XIV. After his marriage to Mary Morland, a collector and illustrator of fossils, the couple spent their honeymoon on a one-year tour of the main museums and sites in Europe. But before settling down to paleontology and married life (the Bucklands had nine children), he took holy orders. He wanted to reconcile the biblical account with the geological and fossil records. At a time when Ducrotay de Blainville in France was dismantling some of the big vertebrates that his former master, the late Cuvier, had assembled to demonstrate the extinction and stability of species, Buckland took a stand as his last major defender. The

Flood was a historical fact, the last of the major natural catastrophes that the earth had suffered, all of them followed by a
new creation.[91]

At this juncture a peculiarly British event occurred. A rich and
eccentric nobleman, the Count of Bridgewater, wanted to devote
his wealth to the glory of God and science. Inspired by William
Paley's *Natural Theology* (1802), he left a considerable fortune at
his death in 1829 to launch a vast publishing venture. It was intended to demonstrate how all living organisms, geology, physics,
astronomy—in short, every area of science—harmonized with
natural theology. The sixth of the eight volumes of the *Bridgewater Treatises* was entrusted to Buckland, who produced *Geology and Mineralogy Considered with Reference to Natural Theology* (1836), a bestseller at the time.[92] Buckland's counterattack
on Blainville was a restatement of the naturalist model of Cuvier,
the divine conditions of existence, and the supreme intelligence
of the creation, only achievable by divinity. The example on
which he built a large part of his argument was our fabulous giant
sloth / armadillo. Its broad shoulders, which had suffered the
outraged honor of a nation and been weighed down by the destiny of a republic, the natural reputation of a continent, and
the reputed radical otherness of the lost world, now had to support an even greater responsibility, or at least a more transcen-
dent one: to prove that divine providence was behind creation,
to demonstrate the existence of God.

In the eyes of Buckland, the powerful muzzle of the Megatherium was used to grind roots, its small trunk was to extract
them, its paws were sharp digging instruments capable of
reaching great depths. Its robust hind limbs and the strength of
the tail could be explained by the need for stability and func-
tionality: while it was digging the soil with one of its front paws,

it had to support itself with the other one, its two rear limbs, and its tail. The Megatherium was becoming a sedentary creature. It would not take long to get it on its feet again. Everything was in line with natural theology. The perfect coherence of this lumbering beast manifested the glory of the Lord and the perfection of life on the planet:

> I select Megatherium, because it affords an example of most extraordinary deviations, and of egregious apparent monstrosity; viz the case of a gigantic animal exceeding the largest Rhinocerus in bulk and to which the nearest approximations that occur in the living world, are found in the not less anomalous genera of Sloth, Armadillo and Chlamyphorus.[93]

The Megatherium played the same role as monsters in the early modern era: it demonstrated the plans of providence. Buckland reproduced the images of Pander and d'Alton, already completely divorced from their original context, because they suited his new purpose (Figure 48).

Buckland did not rule out the possibility that the animal had been armored. Perhaps its hind legs, stout ribcage, and the lumbar muscles it was supposed to have possessed were designed to support the weight of a heavy suit of armor. Various discoveries had been made in 1832 in the province of Buenos Aires and the Río Salado (like the Luján, a tributary of the Río de la Plata). British diplomacy and science were reorganizing the traffic in natural products and fossils from Argentina and the New World after their emancipation to have them sent to London instead of Paris.[94] Among these remains were bones and other fragments of Megatheriums, but also osseous tessellated

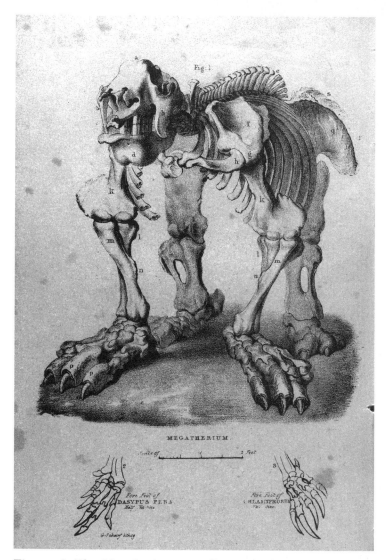

MEGATHERIUM.

Figure 48. The Megatherium according to J. d'Alton, in William Buckland, *Geology and Mineralogy Considered with Reference to Natural Theology* Vol. 2 (1836).

carapaces and even a larger piece of what was clearly a shell like that of a tortoise or armadillo, but gigantic in size. Woodbine Parish, a diplomat and merchant who had lived in Buenos Aires, transported valuable fossil remains to London, including those of a Megatherium, which he presented to the Royal College of Surgeons. The same specimen was used by William Clift (1775–1849), the artist, conservator, and naturalist adopted by John Hunter as his legitimate successor, to present the Geological Society with another of the classic works on the extinct creature, the "Notice on the Megatherium."[95] As Richard Owen later remarked:

> Although, with the characteristic caution of the author, the armour is not directly affirmed to belong to the Megatherium, nothing is stated to prevent the inference that it formed part of the "Remains" of that animal which it is the object of the memoir to describe.[96]

The plates in Clift's article, like some of those that Buckland was to publish in his Bridgewater treatise in the following year, included a small armadillo, whose feet were very similar to those of the Megatherium (Figure 49). Did the Megatherium have the armadillo's armor as well?

Cuvier was succeeded in the chair of comparative anatomy in the Museum of Natural History in Paris by Blainville (his former disciple, who had become his fierce opponent; academic life is also a copy of itself). In 1839, Blainville openly defended the thesis that the Megatherium had been endowed with an impressive armor in the opening of his *Ostéographie*, a long series of iconographic descriptions comparing fossils with living mammals.[97] Like Buckland, Blainville followed anatomical

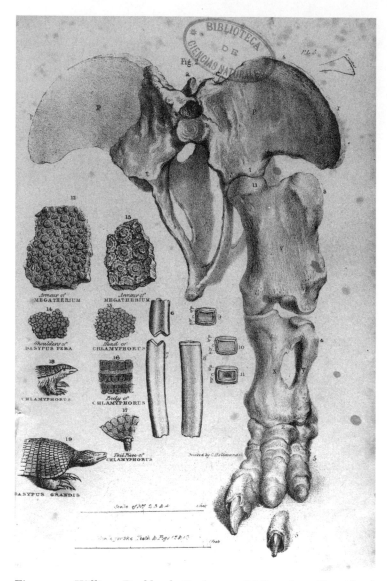

Figure 49. William Buckland, *Geology and Mineralogy Considered with Reference to Natural Theology* Vol.2 (1836).

reasoning: he was confused by the angle of the ribs and the articulation of the unusual pelvis with the vertebral column.

Not until the arrival on the scene of Richard Owen (1804–1892), the successor to William Clift in the Royal College of Surgeons and one of the great anatomists and paleontologists of all time, was the confusion sorted out and each species assigned its corresponding attribute: armor for the Glyptodons, fur for the Megatheria and Mylodons.[98] With the classification of the extinct edentates *(Xenarthra)*, each one acquired its skin once and for all.

Paradoxically, the British Cuvier ended up by introducing ideas from German transformism into England. The identification and name of the giant armadillo, Glyptodon (literally "stone-tooth"), was an act of homage to the method of the father of comparative anatomy, but Owen's notions on the archetype and the homologies between vertebrates owed much to Von Baer, Pander, and d'Alton (and even to Saint-Hilaire). A student of the collections of large mammal fossils that the *Beagle* brought back from its voyage around the globe (1831–1836), Owen dedicated a series of five articles to the Megatherium in the *Philosophical Transactions* (1851–1859), in which he more or less definitively fixed the species.[99]

Owen's systematic work left few loose ends. Like Hooke before him, the occupant of the chair of anatomy and physiology of the Royal College of Surgeons used the microscope to examine the pieces from Punta Alta in northern Patagonia. This was where Darwin initially confused the remains of a Megatherium with those of a rhinoceros, as he recorded in his diary entry for September 23, 1832.[100] Leaving the mistaken identification aside, the reference to size was a commonplace. In his account of the voyage of the *Beagle,* after mentioning the remains of the big mammals and trying to impose order on the different species, he declared the great size of their bones "truly wonderful"; the stature of one of them reminded him of a rhinoceros.[101]

What is certain is that the Megatherium never had armor; it had been confused with the remains of some not-so-remote relatives, the Glyptodons (Figure 50), those gigantic armadillos with whom the Megatheriums shared a habitat for some five million years (both species lived from the Pliocene down to their extinction about nine thousand years ago—a mere moment in geological terms). In 1842, soon after identifying and naming the Glyptodon, the true owner of the armor that had been wrongly attributed to the Megatherium, Owen drew what was destined to become a famous name from his stock of nomenclature when he sought to label the large reptiles of the Mesozoic: he called them *Dinosauria*.[102]

Back in 1818, Buckland had shown Cuvier the remains of a Jurassic animal found in Stonesfield in Oxfordshire that were not very different from the ones that Cuvier had seen years earlier in Honfleur. The godfather of the Megatherium supposed the existence of big antediluvian reptiles, but it was Buckland who informed the Geological Society of London in 1824 that the sharp serrated teeth could not belong to a mammal, but must be those of a reptile. He named the giant carnivorous lizard Megalosaurus. In the following year, Gideon Mantell (who together with Buckland and Owen formed the triumvirate of dinomania) named another colossal reptile, though an herbivorous one this time, Iguanodon because of the resemblance of its teeth to those of the iguana (Figure 51). Twelve-year-old Mary Anning found its remains, the brilliant and often ignored precursor of the great paleontological discoveries that were fast

Figure 50. Glyptodon, Heinrich Harder (1858–1935) *(above)*; Glyptodon, *Museo civico di Storia Naturale*. Milan, photograph of Giovanni-dall'Orto *(below)*.

Figure 51. The Iguanodon dinner party, New Year's Eve, 1853. Notice the plates with names of Buckland, Cuvier, Owen, and Mantell. *The Illustrated Encyclopaedia of Dinosaurs* (1854).

approaching. In 1811, during one of her explorations of the beach near her home of Lyme Regis on the south coast of England (a stretch popularly known today as the Jurassic Coast), the young girl had unearthed the first of a species that would later be called Ichthyosaurus ("fish-lizard").

The big edentates were from this point overshadowed by these protagonists of the resurrection of the antediluvian world, those extraordinary creatures that formed the apex of millenniums of legends of chimeras. Dinosaurs were catapulted to fame by the long tradition of fantastic zoology for many of the same reasons as the rhinoceros and the Megatherium: their strange morphology, their hybrid character, and their remote origins (value was always associated with rarity and novelty). There can be no denying that their magnetism was also due to their effect on the imagination, a powerful force in the collective unconscious of different cultures, from the humanist courts and Napoleonic Europe, the emerging New World of America, or the Victorian era, to our own times. The reasons must be sought in some mysterious fold of the cerebral cortex, but the fascination that reptiles, wildlife, evil, a ferocious world, and stark animality have always provoked is beyond doubt. We have to give it form.

Our journey approaches its end with Owen, who stabilized the anatomy of the Megatherium and clarified its habits, that is to say, who finished the undertaking of Cuvier: by giving it a skin and a history, an identity and habits—a life. Owen definitively ruled out the possibility of its having armor or having lived suspended from the trees. Instead of climbing them, what the Megatherium probably did was to rise up on its two strong hind legs and tear off leaves for food with its front claws (Figure 52). The secret was in the pelvis. The one in Madrid was in a fairly bad condition, and it was only when new specimens arrived in Paris and London that it became possible to reconstruct it properly. The wing-shaped bones of the ilium of the Megatherium (the only part of the hip that did reach Madrid, in spite of everything) have a curved, very concave shape like a fan, which is related to its ability to raise itself on its two sturdy rear legs and to maintain an erect position.

The Megatherium was a digger of roots that adopted a different posture when it had to reach food among the branches of trees. Its apparent sloth and heaviness were disappearing. Its anatomical characteristics were perfectly suited to the kind of life it lived in the South American forests and pampas. This is explained not by divine providence but by its lengthy survival in the region from the Pliocene down to the early Holocene (the geological epoch in which we are situated). The big quadruped was becoming an (occasional) biped, an ingenious solution in

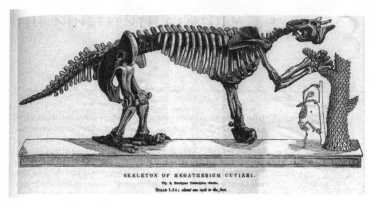

SKELETON OF MEGATHERIUM CUVIERI.
Fig. 2. Skeletus Totocurphus chinois.
Scale 1-14; about one inch to the foot.

Figure 52. Henry A. Ward, *Notice of the Megatherium Cuvieri* (1864).

view of the difficulties that its nonungulate limbs (its claws) presented when it came to locomotion.

It had exploited its versatility to the limit.[103] It had been figured and articulated in a variety of ways. Clinging to the trees, it was given the *modus vivendi* of the sloth, its closest living analogue. Underground, it was closer to the armadillo, the mole, the anteater, or the pangolin. At first it had been reconstructed and represented as a large (herbivorous or carnivorous) quadruped, but in the 1830s its anatomy displayed some characteristics that required it occasionally to rear up on its hind legs, which were longer than the front ones, just as bears raise themselves to scrape the bark of trees. This is how the Megatherium is represented today in museums of Natural History in London, Paris, and elsewhere (Figure 53). Presenting it in this position is no doubt also due to scenographic considerations—an erect animal is always more spectacular and terrifying. Still, nothing is eternal: some documentaries show it moving on all fours; someday we will see it run.

Figure 53. Megatherium at the Natural History Museum, London.

In reviewing the case, Darwin recalled that it had been a bold and rather absurd idea to suppose that antediluvian trees could have sustained animals the size of an elephant clambering around in their branches. He should have added that this was a presentist projection: the most natural, immediate, and frequent analogy is to suppose that everything in the past was just like it is now (that the giant sloths of the past lived dangling from equally gigantic trees). Indeed, as he formulated it, "the habits of life of these animals were a complete puzzle to naturalists, until Professor Owen lately solved the problem with remarkable ingenuity."[104] The assembly and physical reconstruction, as well as the reconstruction of its habits of life, had been as disconcerting as they had been tantalizing, for anatomy and life are as inextricably intertwined as morphology and history.

EPILOGUE

Circular Lives

For unintentionally, each one desires the opposite of himself
in order to have the whole.

Goethe, *The Metamorphosis of Plants*

This essay is drawing to a close. Like experimental scientists, we have manipulated the phenomena and exposed them to artificial conditions. We have interrogated, observed, and described them. They have said what they had to say, or at least what we have been able to get them to say. There are no further surprises in store. The game is over.

More than parallel lives, the trajectories of the rhinoceros and the Megatherium offer a curious case of circular biographies (Figure 54). There is something enveloping, repetitive, and symmetrical about their movements, concentric circles that can merge or separate. The two big vertebrates traveled around the world at two strategic moments of the modern era. One of them sailed halfway around the world, rounding the Cape of Good Hope just a few years before the first circumnavigation of the globe. That feat had been foretold by Behaim's terrestrial globe, that marvelous object made in Nuremberg, the city of Dürer. The hoofed pachyderm arrived in Lisbon in 1515. Thanks to Dürer's woodcut, the German presses soon

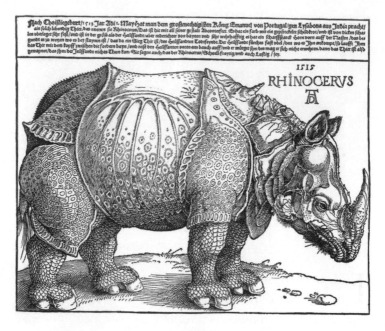

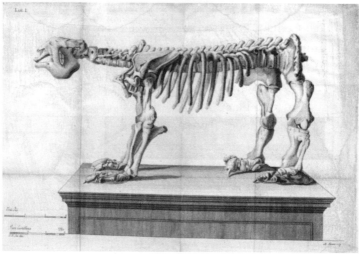

Figure 54. The two primary images of this book: Dürer's woodcut and Bru and Navarro's engraving.

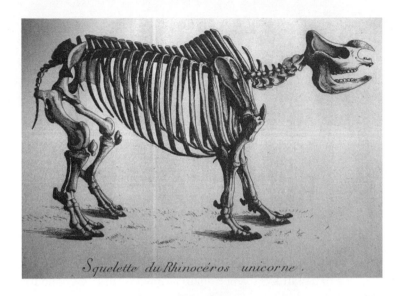

Squelette du Rhinocéros unicorne.

Figure 55. The "negative" or the reverse of the previous images: the skeleton of the rhinoceros drawn by Cuvier, in *Recherches* (1812), and the Megatherium with its fur, drawn by Mauricio Antón (courtesy of the artist).

disseminated its image in all directions. The image traveled through space and time across Europe and right down to the Enlightenment. It even reached America and returned to colonize ideas about the Orient. Few images better express what the first globalization meant.

The Megatherium reached Europe from where the sun sets, following the Copernican course that history took once Europeans had stopped gazing at their origins and set out to conquer the future. The Atlantic and the New World are the undeniable protagonists of the modern world. It seems appropriate (in terms of symmetry and order—a concept that the Greeks called *kosmos*) that our second large vertebrate, burdened with history, should come from that world of the future. It traveled to Madrid from the Río de la Plata in 1788. Once again a drawing and an engraving permitted its image to circulate, first to Paris and from there around the world. Few images better capture what happened after the closing of the global circle: the emergence of (deep) time. Shortly after Cook had completed the spherical task begun by Magellan, Laplace concluded Newtonian-based celestial mechanics. Now that the earth had been circumnavigated and removed from its place at the center of the cosmos, the human race was confronted with astronomical dimensions. Hutton, Lyell, Lamarck, and Darwin revolutionized the history of the earth and of life; Humboldt produced his *Kosmos*. The geographical circle or sphere was closing, and the last unknown spaces were being mapped, just at the moment when a spiral of knowledge and uncertainty about a world of other matters opened up.[1]

We can assign properties to our two protagonists that make them resemble objects: they are commodities, fetishes, luxury articles, curiosities, cadavers, museum exhibits, engravings. But we can also assign them protohuman aspects, since they were brought to life and played remarkable social roles, became famous, and performed symbolic and political functions in courts, republics, and scientific disciplines. Were they objects of trade? Their high mobility and dynamism seem more appropriate for *subjects* who act as if they had a life of their own. This story has also been the double biography of two creatures that traveled around the globe to acquire what they lacked. The rhinoceros, an amalgam of armor and legend, a fantastic animal associated with the word, the emblem, and allegory, advanced ponderously through the treatises of natural history of the early modern era. Standardized by the art of Dürer, it moved invincibly from page to page in the works of Münster, Gessner, Aldrovandi, Thevet, Paré, and Jonston while remaining identical to itself.

But its strength is our weakness. The *rhinoceros unicornis*, an odd-toed member of the Placentalia subdivision of mammals, was to remain screened by that image that dictated its life and properties. From Dürer on, artists and scholars concentrated on its outward appearance, above all its formidably tough hide, or armor as Dürer would have it. This outward focus and Ganda's singularity—once he was lost at sea, there was no other exemplar to examine—meant that the internal organs and physiology of the rhinoceros remained *terra incognita*. We have seen how early modern science encircled and measured the surface of the earth at the same time that it dissected and opened up bodies of other creatures. This is what the life sciences would be dedicated to: observing, describing, and illustrating the interior of organisms. From Vesalius onward, no task was more urgent than to disclose hidden anatomy, explore the interior of things, visualize authentic nature beyond appearance, beneath the surface, behind skin and legend, where unknown phenomena lay in wait to be revealed and diffused, those hidden things that sooner or later surface and come into view. They are the most prosaic and mundane, but they support and articulate the world and give life a spinal column: bones.

Who will reveal their osteology? Who will penetrate the layer of skin to x-ray its underlying form, to illustrate its bones? The answer is Cuvier, the same man who bestowed a missing skin on the Megatherium. Among his many works on osteology we find the singular image of this rhinoceros finally x-rayed, stripped of its legendary armor, of Pliny and Dürer, just as it is (Figure 55).

Exactly the opposite happened to the strange skeleton from the River Luján. That heap of mute and scattered bones required the words of a Pliny or the hand of a Dürer, a taxidermist, or a taxonomist, to assemble and arrange it—to bring it to life, to provide it with a skin that would give it a true anatomical form and all that this entails: forms of nutrition, movement, and behavior, some character traits. What should we call all of this? A physical manifestation with a narrative? A life? Perhaps, simply, a history. The Megatherium was an antediluvian creature eager to regain what it lost thousands of years earlier.

The rhinoceros and the Megatherium both came back to life in an image. Their journeys can be understood metaphorically as the search for what they lacked, for fragments of their anatomy and identity.

The central analogy that we have established between them both is the inverse correspondence that they have with skin and bones, the most external and the most internal parts of the anatomy. The rhinoceros is associated with the former, the Megatherium with the latter. The rhinoceros is hide, armor, and horn; that is, it has an identity, a name, a legend, habits, moral values, biblical resonances, a considerable humanist and classical baggage associated with what is most readily seen. It also has an end or purpose: to confront the elephant, to symbolize the struggle between two colossal, opposing forces. Its horn is supposed to pierce the belly of its rival, the most popular (and the most human) among the big mammals. Its duel in the Terreiro do Paço was a counterexperiment staged under the dictates of ancient natural history: actions had to be adjusted to fit words. In Dürer's woodcut, the legend predominates, not the image. The rhinoceros is what Strabo and Pliny had said that it was, or should be. Words conditioned the image. Facts were subordinated to a preordained appearance. Its life was prescribed, its form was prefigured.

The Megatherium, on the other hand, began as the silent, scattered bones of something nameless and unknown. While the disembarking of Ganda in Lisbon caused an uproar, the skeleton from the River Luján arrived as silently as the grave, segmented, without any established attributes, surrounded in a halo of mystery. Nobody knew what it was. Midway between forensic anatomy and vertebrate paleontology, the case began to arouse the interest of the scientific community. After spending some time in crates, the skeleton was sketched and assembled in the Royal Cabinet of Natural History in Madrid. These were the images that attracted the words, attributes, and predicates intended to flesh it out. Taxonomy and taxidermy were about putting it in its place and giving it a skin. Various options were tried out until its place could be stabilized and fixed in a hitherto unoccupied location in the history of life. It took a while, but in the end it managed to regain its skin and its ancient life. The history of this vertebrate fossil is that of a hidden phenomenon that rises to the surface, invisible until its discovery and disinterment. It was revolutionary too: the knowledge that was derived from it did not confirm the status quo of written knowledge or of the world. It tended, rather, to undermine them.[2]

There are other differences between the two big vertebrates. The formal stability that our armored unicorn acquired in the woodblock of the master of Nuremberg contrasts with the flexibility of the ungulate herbivore with its protean propensity for metamorphosis. The creature of the East arrived alive in Europe and was immortalized by art and science through the effigy of an unalterable image. The beast from the West was dead on arrival and was resuscitated by art and science. These two opposite miracles both demonstrate the power of images to elude or outwit death, to make things animate.

This essay has explored how science and art worked on the two subjects: both the scientific life of the image of the rhinoceros and the character of the image of the Megatherium as a zoological *vanitas*. It is also intended to be the double history of two images that not only managed to immortalize and resuscitate (globalize or universalize) two natural species, but did so in spite of the doctrine of direct contact and *ad vivum* presence. Neither Dürer nor Cuvier saw the exemplars that they nevertheless managed to immortalize or reincarnate. The experience *ex visu* took place through an intermediary: proxy pictures, delegated drawings. Such examples of indirect witness and delegated knowledge abound, given the collective nature of scientific practice. The imagination, the capacity to produce images that can be perceived by the senses and that materialize the ideal, is a faculty that often depends on distance and abstraction, the free association of data, features, and forms. To produce knowledge it can help to detach yourself from events. For instance, once the Megatherium had been sketched and represented in scale—in other words, once it became easier to discount its alarming dimensions—it could be seen as a sloth. It is easier to perceive the analogy between the two species if we detach ourselves from a fact that is as obvious as it is misleading: their different sizes.

Paradoxically, we have to close our eyes, blot out certain factors, and construct our images not from what is there, but from what ought to be or that we hope and expect to be there. Such an ability to visualize explains the presence in Dürer's engraving of details and forms deriving from an imaginary bestiary. The imagination is the main tool we use to explore the possibility of being and to give form to unprecedented or invisible things, whether they be a matter of fable (the Leviathan or the Minotaur) or the opposite (the Megatherium or natural selection, neither visible nor capable of being observed simply with the eye).

If the imagination serves to bring things that are distant in space and time closer, the images were the means of transport used by our two travelers. The example of the engraving has given us the opportunity to observe the effects of the mechanical production and circulation of knowledge. The universalization of phenomena arises from an artisanal technique that converts them, in the words of Bruno Latour, into immutable mobiles.[3] The copy lies at the heart of the question of knowledge and representation, not only for its multiplying effect and its capacity to produce a community of virtual witnesses—to produce science—but also because it takes us back to the tricky problem of the presence and absence of the phenomenon observed, the object, and also of the subject. To represent a living creature or a natural phenomenon of any kind, to imagine them, there is no need to have them in front of you or to follow those who claim to have seen them directly. Presses and plates work independently of whoever designed the original. The whole edifice of science, whether we like to admit it or not, is based on trusting simulations or depictions of what we have not seen ourselves.

This might seem to put us on shaky ground. For it is impossible to talk about a copy without talking about falsification. The key word in the caption of *Rhinocerus 1515* was *abconterfect*, a polysemic term that encompassed both the duplicity of the fidelity of the *ad vivum* portrait and fraudulent imitation. Every image assumes the same ambiguity: it is an adulterated, manipulated image, an idol. The two primary images discussed in this essay, those by Dürer and Bru, open up a spectrum of problems connected with this fact. Dürer's rhinoceros does so because it inaugurated the era of mechanical reproduction,[4] which allowed the image to enter the home of any merchant or artisan in the sixteenth century and endure right down to the museum shops of today. As for the image of the Megatherium, it raises an epistemologically relevant issue: How can you produce a copy, an imitation, without anyone at all having seen the original? How do you articulate or compose an image of an extinct animal when there is no model? Sometimes forging an image of things, making it possible to see them, requires really imagining them.

Fossils, moreover, offer another possibility in this sense, a further chapter in the dialectic between art and nature. For centuries the question was: Are fossils authentic creatures and natural forms, or just copies arising from the plastic properties of the material that forms them? Might they be natural artifices? Were they original or duplicates? The history of how these singular, accumulated, and treasured products—as rare and curious as exotic fauna—were interpreted has plunged us into the prehistory of paleontology, an era of scientific hypotheses destined to try to reply to questions that resonated with those that underlie the rest of this essay. How were the natural phenomena to be squared with words (or in this case, with Holy Scripture)? How could you give form to something so distant

Figure 56. The Megatherium from the Río Luján.

and hard to imagine as life thousands, millions of years ago? How can you produce or reproduce what is different?

Even more extraordinary than the rhinoceros was for the Renaissance courts, the Megatherium came from a country far from the salons of the Enlightenment, from a time of which there was no human testimony, a totally foreign past that needed to be given form. The transfers between archaeology and paleontology reflected an analogy that we have frequently encountered in this essay: that between scientists and historians, who always learn from one another when it comes to dealing with and manipulating the data of nature and of the past. The students of fossils concentrated on the patient and apparently modest task that the antiquarians performed with material remains, in the same way that Carlyle, for example, was to end up applying the similes of Victorian paleontology to explain his own historical method.[5] This crossing of perspectives, the use of borrowed standpoints on the margins of a given discipline, has been one of the most constant and visible phenomena in the history of knowledge. Why not resort to that? Why not try to learn from our object of study to think about how we historians act and produce our knowledge?

This has been an essay, not a treatise. It is self-reflexive insofar as our practice is self-reflexive. It is unstable, tentative, and personal inasmuch as any essay, from Montaigne onward, works with a fluid material that resists objectification.

The aim has been to find the isomorphic lines between two species and two histories that are very different on the surface. The rhinoceros and the Megatherium have in common their role as vestiges or residues of remote realities. One of them brought the marvels of the Orient, the other a deep and foreign time. They both have something extraterrestrial about them; their function was to expand the margins of what was real and possible, definitively, of the imagination. When all is said and done, they were both extraordinary creatures. Their monstrosity and hybrid character put them on an equal footing with chimeras, prodigies of nature drawn from some compendium of natural wonders or manual of fantastic zoology, that universal book that includes life from the unicorn to the dinosaur, the sum total of creatures that have been and that might have been.

Is it possible to establish analogies or homologies between such disparate forms of life and historical periods? We have made the effort. In spite of the difficulty of understanding the world and the irreducible nature of the tangible, in spite of the obstinate presence of the local and the singular, some of us continue to believe in the affinity of processes of knowledge, in the continuity of history, in the profound identity of life, beyond the laws of variation and beneath the apparent metamorphosis of things.

Notes
Acknowledgments
Credits
Index

NOTES

Prologue

1. As elsewhere, there have been various public debates in Spain on the recovery of the past, memory, and forgetting. When the Spanish version of this book was being completed in 2009, an intense debate was under way over what to do with the traumatic past of the Spanish Civil War.

2. Martin Kemp, *Behind the Picture: Art and Evidence in the Italian Renaissance* (New Haven: Yale University Press, 1997), pp. 237ff.

3. María Zambrano, *La agonía de Europa* (Madrid: Mondadori, 1988), p. 11.

4. Translated into English as *The Grammar of Fantasy: An Introduction to the Art of Inventing Stories* (New York: Teachers and Writers Collaborative, 1996).

5. Florike Egmond and Peter Mason, *The Mammoth and the Mouse: Microhistory and Morphology* (Baltimore: Johns Hopkins University Press, 1997), p. 39.

6. Lawrence Norfolk, *The Pope's Rhinoceros* (London: Sinclair-Stevenson, 1996).

7. W. J. T. Mitchell, *Iconology: Image, Text, Ideology* (Chicago: University of Chicago Press, 1986), p. 43.

8. Egmond and Mason, *The Mammoth and the Mouse*, p. 33.

9. Steven Shapin, *Never Pure, Historical Studies of Science as if It Was Produced by People with Bodies, Situated in Time, Space, Culture, and Society, and Struggling for Credibility and Authority* (Baltimore: Johns Hopkins University Press, 2010).

10. Mary Douglas, *Purity and Danger: An Analysis of Concepts of Pollution and Taboo* (London: Routledge and Keegan Paul, 1966).

1. Itinerary

1. Ganda's journey to Europe was recorded at the time by various chroniclers and humanists (Damião de Gois, Gaspar Correia, Diogo do Couto, João de Barros, Paolo Giovio, Valentim Fernandes) and has

attracted the attention of a variety of historians since then (Bernardo Gomes de Brito, Angelo de Gubernatis, Abel Fontoura da Costa, Silvio A. Bedini, T. H. Clarke, Donald F. Lach, Ugo Serani). It is even the theme of the novel *The Pope's Rhinoceros* by Lawrence Norfolk (London: Sinclair-Stevenson, 1996).

2. Isabel Soler, ed., *Los mares náufragos* (Barcelona: El Acantilado, 2004), p. 13.

3. Mauricio Jalón, "Los viajes y las ciencias al inicio de la modernidad," in *Andanzas y Caminos: Viejos libros de viajes*, exhibit catalogue, Museo de Pasión, Valladolid (Valladolid: Junta de Castilla y León, 2004), pp. 109–127, here p. 117.

4. The classic work on the subject is Marcel Mauss, "Essai sur le don: Forme et raison de l'échange dans les sociétés archaïques," *Année Sociologique*, n.s. 1 (1923–1924): 30–186. English translation: *The Gift: The Form and Reason for Exchange in Archaic Societies*, trans. W. D. Halls (New York: W. W. Norton, 1990).

5. Abel Fontoura da Costa, *Les déambulations du Rhinocéros de Modofar, roi de Cambaye, de 1514 à 1516* (Lisbon: Agencia Geral das Colónias, 1937), p. 10.

6. Edward W. Said, *Orientalism* (New York: Pantheon, 1978).

7. The metaphor is taken from Isabel Soler, *El nudo y la esfera: El navegante como artífice del mundo moderno* (Barcelona: El Acantilado, 2003).

8. Auguste Toussaint, *Historia del Océano Indico* (Mexico City: Fondo de Cultura Económica, 1984), p. 44ff.; Charles Boxer, *The Portuguese Seaborne Empire, 1415–1825* (New York: Alfred A. Knopf, 1969).

9. Gaspar Correia, *Lendas da India* (Lisbon, 1858–1864), vol. 2, pp. 373–374.

10. Toussaint, *Historia del Océano Indico*, p. 48. Kapil Raj, in *Relocating Modern Science: Circulation and the Construction of Knowledge in South Asia and Europe, 1650–1900* (Basingstoke, UK: Palgrave, 2007), also stresses the excesses of British and by extension European historiography in exaggerating the role of the activity of westerners in India, while in fact they were just one among numerous actors involved in a complex and diverse cultural space.

11. Eric Baratay and Elisabeth Hardouin-Fugier, *Zoo: A History of Zoological Gardens in the West* (London: Reaktion Books, 2002), pp. 17–28.

12. Daniel Hahn, *The Tower Menagerie: The Amazing 600-Year History of the Royal Collection of Wild and Ferocious Beasts Kept at the Tower of London* (London: Simon & Schuster, 2003).

13. On natural history collections in Renaissance Italy, see Paula Findlen, *Possessing Nature: Museums, Collecting, and Scientific Culture in Early Modern Italy* (Berkeley: University of California Press, 1994).

14. On the Medici giraffe, see Marina Belozerskaya, *The Medici Giraffe and Other Tales of Exotic Animals and Power* (New York: Little, Brown, 2006), pp. 87–131.

15. Pliny, *Natural History* (Madrid: Cátedra, 2002), 8.27, p. 86.

16. Arjun Appadurai, ed., *The Social Life of Things: Commodities in Cultural Perspective* (Cambridge: Cambridge University Press, 1986).

17. Lorraine Daston, ed., *Biographies of Scientific Objects* (Chicago: University of Chicago Press, 1999); Lorraine Daston, ed., *Things That Talk: Object Lessons from Art and Science* (New York: Zone Books, 2004).

18. Georg Simmel, *Philosophy of Money* (1900), cited in Appadurai, *The Social Life of Things*, p. 3.

19. Silvio A. Bedini, "The Papal Pachyderms," *Proceedings of the American Philosophical Society* 125, no. 2 (April 30, 1981): 75–90.

20. Fontoura da Costa, *Les déambulations du Rhinocéros de Modofar*, p. 29.

21. Ibid., p. 28.

22. Bedini, "The Papal Pachyderms"; Silvio A. Bedini, *The Pope's Elephant: An Elephant's Journey from Deep in India to the Heart of Rome* (Manchester, UK: Carcanet Press, 1997).

23. Garcia de Resende described the mission in his *Crónica dos Feitos de D. João e Miscelânea*. I owe this information and several other details of this episode to an unpublished manuscript by Isabel Soler, "Viaje y humanismo en Portugal," with thanks to the author for sharing it with me.

24. Bedini, "The Papal Pachyderms," p. 77.

25. Derived from the Hindu original *Ana*, given an Italian touch in the extended form *Annone*, abbreviated to *Anno*, and finally appropriated through the prevalence of the English language as *Hanno*, the history of this single word says it all. Hanno was the name of the Carthaginian general of Hannibal's army that crossed the Strait of Gibraltar with its elephants, and also of a famous lion tamer in ancient Rome. The history of a single word also tends to confuse everything.

26. Bedini, "The Papal Pachyderms," p. 80.

27. Ibid., p. 83.
28. See *História Trágico-Marítima* (1735–1736), edited by Bernardo Gomes de Brito, some of whose most important texts have been edited by Isabel Soler in *Los mares náufragos*.

2. Words

1. Like all the other historians who have written about the episode, we follow two main sources: the 1556 edition of Damião de Góis, *Crónica do Felicíssimo Rei D. Manuel* (Coimbra, Portugal: Universidade de Coimbra, 1955), vol. 4, chap. 18, pp. 49–55; and Valentim Fernandes, *Lettera scripta de Valentino Moravio, Germano, a li mercantanti di Norimberga*, reproduced in Abel Fontoura da Costa, *Les déambulations du Rhinoceros de Modofar, roi de Cambaye, de 1514 à 1516* (Lisbon: Agencia Geral das Colónias, 1937), pp. 33–41.
2. Bernardo Gomes de Brito, "Os pachydermes do Estado d'El-Rei D. Manuel," in *Revista de Educação e Ensino* (1894), pp. 79–86.
3. On the city of Lisbon as a hot spot of the international trade in commodities and wild animals, see Annemarie Jordan and K. J. P. Lowe, eds., *The Global City: On the Streets of Renaissance Lisbon* (London: Paul Holberton, 2015).
4. Góis, *Crónica do Felicíssimo Rei D. Manuel*, vol. 4, p. 49.
5. Ibid., p. 53.
6. Ibid., p. 54.
7. Clifford Geertz, "Deep Play: Notes on the Balinese Cockfight," in *The Interpretation of Cultures* (New York: Basic Books, 1973).
8. "Se viene a Roma e là potré far male, / ma noi faremo in modo se ci viene / che ciaschedun sarà bestia da bene"; Giovanni Giacomo Penni, *Forma e Natura e Costumi de lo Rinocerote stato condutto in Portogallo dal Capitanio de l'Armata del Re et altre belle cose condutte dalle Insule novamente trovate*, reprinted with Spanish translation in Ugo Serani, "Forma e natura e costumi de lo rinocerote de Giovanni Giacomo Penni," *Etiópicas* 2 (2006): 146–171. The poem was printed in Rome in the house of the master Stephano Guilireti and is dated July 13, 1515. A copy once belonging to Hernando Colón can be found in the Biblioteca Colombina in Seville. Penni's pamphlet was very probably a replica of an earlier one entitled *Forma e natura e costumi de lo elefante* and dedicated to Hanno, the pope's elephant.

9. "Aciò che meglio possin dar diletto / varie altre cose per viso e per mane / et animali che fanno odor nel letto / e tante gentileze e tante frasche / che ciascuna potrà impir le tasche."

10. The reference is to the last book of Northrop Frye, *Words with Power: Being a Second Study of the "Bible and Literature"* (Toronto: University of Toronto Press, 1990).

11. Anthony Grafton, *Defenders of the Text: The Traditions of Scholarship In an Age of Science, 1450–1800* (Cambridge: Harvard University Press, 1991).

12. Umberto Eco, *The Search for the Perfect Language*, trans. James Fentress (Oxford: Blackwell, 1995).

13. Lorenzo Valla, *De elegantiis latinae linguae* (1471).

14. Góis, *Crónica do Felicíssimo Rei D. Manuel*, p. 49.

15. Pliny, *Natural History* (Madrid: Cátedra, 2002), 8.29, p. 86.

16. Ibid., 1–11.

17. Ibid., 29. Pliny's rhinoceros is an Indian one; the African rhinoceros has two horns but is less long-lived and robust.

18. Ibid., 11, 12.

19. The other scattered references to the rhinoceros in Pliny add little substance.

20. Strabo's work, composed between 29 and 7 BC, comprises seventeen books. This encyclopedic work, replete with stories, myths, knowledge, and beliefs about the various peoples of the ancient world, marked a change of direction in the mathematical geography of Eratosthenes and Hipparchus of Nicaea. The indefatigable traveler reached Armenia, Ethiopia, and the Black Sea, and included many other regions on the basis of various testimonies and accounts.

21. Strabo, *Geographia*, 16:4, 15.

22. William Ashworth Jr., "Natural History and the Emblematic World View," in David C. Lindberg and Robert S. Westman, eds., *Reappraisals of the Scientific Revolution* (Cambridge: Cambridge University Press, 1990), pp. 303–333; William Ashworth Jr., "Emblematic Natural History of the Renaissance," in N. Jardine, J. A. Secord, and E. C. Spary, eds., *Cultures of Natural History* (Cambridge: Cambridge University Press, 1996), pp. 17–38; Paula Findlen, *Possessing Nature: Museums, Collecting, and Scientific Culture in Early Modern Italy* (Berkeley: University of California Press, 1994); Brian W. Ogilvie, *The*

Science of Describing: Natural History in Renaissance Europe (Chicago: University of Chicago Press, 2006).

23. On curiosity in the early modern period, see Justin Stagl, *A History of Curiosity: The Theory of Travel, 1550–1800* (Chur, The Netherlands: Harwood, 1995); Lorraine Daston and Catharine Park, *Wonders and the Order of Nature, 1150–1750* (New York: Zone Books, 1998).

24. Peter Mason, *Before Disenchantment: Images of Exotic Animals and Plants in the Early Modern World* (London: Reaktion Books, 2009).

25. Solinus's compilation, known as *De mirabilibus mundi* or *Collectanea Rerum Memorabilium*, was published in 1895 under the latter title by Theodor Mommsen, who dated the work to the third century AD.

26. On American natural history in the first age of expansion, see José Pardo Tomás, *Oviedo, Monardes, Hernández: El tesoro natural de América: Colonialismo y ciencia en el siglo XVI* (Madrid: Nivola, 2002).

27. Italo Calvino, *Why Read the Classics?*, trans. Martin McLaughlin (New York: Pantheon Books, 1999).

28. Edward W. Said, *Orientalism* (New York: Pantheon, 1978); Jacques Le Goff, *The Medieval Imagination*, trans. Arthur Goldhammer (Chicago: University of Chicago Press, 1988).

29. Francisco Javier Gómez Espelosín, *El descubrimiento del mundo: Geografía y viajeros en la Antigua Grecia* (Madrid: Akal, 2000), p. 255.

30. There is a chapter on the relation between the two one-horned animals (the unicorn and the rhinoceros) in Bruno Faidutti, "Images et connaissance de la licorne: Fin du Moyen-Age—XIXème siècle" (Ph.D. thesis, Université Paris XII, Sciences littéraires et humaines, 1996), http://faidutti.free.fr/licornes/these/these.html. Faidutti presents a detailed account of the tortuous history of distinctions and confusions between the two animals that is more persistent than it might appear. The engravings in two outstanding illustrated natural histories of the sixteenth-century Renaissance, those by Conrad Gessner and Ulisse Aldrovandi, inspired the great illustrator and engraver Antonio Tempesta. In 1636 he published a series on quadrupeds that kept up the ambiguity between a *monocerons* with a spiraled horn and a *renocerons* with an equine shape and a pig's tail, identified with the Plinian *monoceros*. Both were given the same

name, *alicornio*, in Italian. Two decades later, in *Historia Naturalis de Quadrupedibus* of 1657, Jan Jonston still used the pleonasm *monoceros unicornu* to identify the rhinoceros whose image was derived from that of Tempesta.

31. Carl Jung, *Psychology and Alchemy*, 2nd ed. (London: Routledge, 1980).

32. The words appear on the tapestry dedicated to the sixth sense, but the other five also include a unicorn.

33. Vol. 1 of Vladimir Acosta, *Viajeros y maravillas* (Caracas: Monte Avila, 1993) is devoted to the various versions of the travels of Alexander the Great in Asia, Africa, and Europe.

34. Ibid., pp. 64–69.

35. John Larner, *Marco Polo and the Discovery of the World* (New Haven: Yale University Press, 1999).

36. Marco Polo, *The Travels*, trans. Ronald Latham (Middlesex: Penguin, 1958).

37. Acosta, *Viajeros y maravillas*, vol. 3, pp. 173–191.

38. Polo, *The Travels*, p. 253.

39. *The Travels of Sir John Mandeville*, trans. C. W. R. D. Moseley (Middlesex: Penguin, 1983).

40. For this and what follows, see Góis, *Crónica do Felicíssimo Rei D. Manuel*, vol. 4, chap. 18.

41. Ibid., p. 50.

42. Ibid., p. 51.

43. Ibid., p. 52.

44. Numbers 23:22 and Job 39:9–12.

45. Góis, *Crónica do Felicíssimo Rei D. Manuel*, p. 53.

46. Ibid., p. 54.

47. Le Goff, *The Medieval Imagination*.

48. The *cour des miracles* was traditionally a quarter of medieval Paris inhabited by prostitutes and beggars. The Spanish reader, however, will recognize the allusion to the title of the 1927 novel *La corte de los milagros* by Ramón del Valle-Inclán, which parodies the vices and eccentricities of the court of Isabel II (1833–1866).

49. As Orpheus called the goddess in Ficino's *Commentary on the Symposium of Plato*. Pedro R. Santidrián, ed., *Humanismo y Renacimiento* (Madrid: Alianza, 1986), p. 59.

3. Print

1. Paolo Giovio mentions the episode in at least two places: in *Dialogo dell'Imprese Militari e Amorose*, 1555 (translated into Spanish by Alonso de Ulloa as *Diálogo de las empresas militares y amorosas* (Lyon: Guillermo Roville, 1562), and in *Elogia virorum bellica virtute illustrium veris imaginibus supposita*, 1561. Abel Fontoura da Costa, *Les déambulations du Rhinoceros de Modofar, roi de Cambaye, de 1514 à 1516* (Lisbon: Agencia Geral das Colónias, 1937), p. 44; Ugo Serani, "Forma e natura e costumi de lo rinocerote de Giovanni Giacomo Penni," *Etiópicas* 2 (2006): 146–171, here p. 153.

2. William Shakespeare, Sonnet 18.

3. On Leonardo and the link between science and art, see Martin Kemp, *Leonardo* (Oxford: Oxford University Press, 2004). The whole of Kemp's oeuvre, and not just his work on Leonardo, is concerned with the relation between science and art.

4. On Dürer's insistence on the distinction and complementarity between theory and practice, knowledge and technique—or, in more encyclopedic terms, "science and art," see Erwin Panofsky, *The Life and Art of Albrecht Dürer*, 4th ed. (Princeton: Princeton University Press, 1955), p. 242.

5. The literature on the relation between science and art in the early modern age and on the role of the image in the production and circulation of knowledge has grown exponentially in the past few decades. See, for example, Allan Ellenius, ed., *The Natural Sciences and the Arts* (Uppsala: S. Academiae Ubsaliensis, 1985); Brian S. Braigie, ed., *Picturing Knowledge: Historical and Philosophical Problems concerning the Use of Art in Science* (Toronto: University of Toronto Press, 1996); Pamela H. Smith and Paula Findlen, eds., *Merchants and Marvels: Commerce, Science, and Art in Early Modern Europe* (New York: Routledge, 2002); Pamela H. Smith, "Art, Science and Visual Culture in Early Modern Europe," *Isis* 97 (2006): 83–100; Sachiko Kusukawa, *Picturing the Book of Nature: Text, Image, and Argument in Sixteenth-Century Human Anatomy and Medical Botany* (Chicago: University of Chicago Press, 2012).

6. John 20:8–9: "Then went in also that other disciple, which came first to the sepulchre, and he saw, and believed. For as yet they knew not the scripture, that he must rise again from the dead."

7. Krzystof Pomian, "Vision and Cognition," in Caroline A. Jones and Peter Galison, eds., *Picturing Science, Producing Art* (New York: Routledge, 1998), pp. 211–232.

8. William Eamon, *Science and the Secrets of Nature: Books of Secrets in Medieval and Early Modern Culture* (Princeton: Princeton University Press, 1994), pp. 269–300 (the chapter is entitled "Science as a *Venatio*").

9. Two books with identical titles but very different approaches appeared in the same year: Irene Baldriga, *L'occhio della lince: I primi lincei tra arte, scienza e collezionismo (1603–1630)* (Rome: Accademia Nazionale dei Lincei, 2002); and David Freedberg, *The Eye of the Lynx, Galileo, His Friends, and the Beginnings of Modern Natural History* (Chicago: University of Chicago Press, 2002).

10. Martin Kemp, *The Science of Art: Optical Themes in Western Art from Brunelleschi to Seurat* (New Haven: Yale University Press, 1990).

11. Panofsky, *The Life and Art of Albrecht Dürer*, p. 279.

12. José Manuel Matilla, ed., *Durero: Obras maestras de la Albertina* (Madrid: Museo Nacional del Prado, 2005), pp. 135–140.

13. Panofsky, *The Life and Art of Albrecht Dürer*, p. 243.

14. Claudia Swan, "*Ad vivum, naer het leven*, from the Life: Defining a Mode of Representation," *Word & Image* 11, no. 4 (1995): 353–372.

15. One of the most influential studies to focus on the convergences between the two processes and on the differences between the model of the Low Countries and that of the Italian Renaissance is Svetlana Alpers, *The Art of Describing: Dutch Art in the Seventeenth Century* (Chicago: University of Chicago Press, 1983).

16. Pamela Smith, "In a Sixteenth-Century Goldsmith's Workshop," in Lissa Roberts, Simon Schaffer, and Peter Dear, eds., *The Mindful Hand: Inquiry and Invention from the Late Renaissance to Early Industrialisation* (Chicago: Edita Amsterdam and University of Chicago Press, 2007), pp. 33–58; Pamela H. Smith, *The Body of the Artisan: Art and Experience in the Scientific Revolution* (Chicago: University of Chicago Press, 2004).

17. On the various aspects and problems connected with the representation of the invisible in the history of art and science, see Martin Kemp, *Seen/Unseen. Art, Science, and Intuition from Leonardo to the Hubble Telescope* (Oxford: Oxford University Press, 2006). The pioneer in this

field is Martin J. S. Rudwick, on whom we draw in the second part of this essay. See his groundbreaking article "The Emergence of a Visual Language for Geological Science, 1760–1840," *History of Science* 14 (1976): 149–195.

18. Lisa Jardine, *Worldly Goods: A New History of the Renaissance* (New York: Norton, 1996), p. 289ff.

19. Ibid., p. 293.

20. Panofsky, *The Life and Art of Albrecht Dürer*, p. 242.

21. The topic of curiosity and exoticism in the early modern period has been discussed by historians of science in numerous places: Lorraine Daston and Katharine Park, *Wonders and the Order of Nature, 1150–1750* (New York: Zone Books, 1998); Barbara M. Benedict, *Curiosity: A Cultural History of Early Modern Inquiry* (Chicago: University of Chicago Press, 2001). It is reflected in the various visual arts in the proliferation of exotic fauna—for example, Leonardo managed to construct a mechanical lion in the same year, 1515, and Titian included a pair of cheetahs soon afterwards in his *Bacchus and Ariadne* (1520–1523). On the representation of exotic fauna, see Miguel de Asúa and Roger French, *A New World of Animals: Early Modern Europeans on the Creatures of Iberian America* (Aldershot: Ashgate, 2005); Peter Mason, *Before Disenchantment: Images of Exotic Animals and Plants in the Early Modern World* (London: Reaktion Books, 2009).

22. Panofsky, *The Life and Art of Albrecht Dürer*. p. 209.

23. Valentim Fernandes, *Lettera scripta de Valentino Moravio, Germano, a li mercantanti di Norimberga*, Biblioteca Nazionale Centrale di Firenze, Code Strozziano, no. 20. Cote: ora CI.-XIII 80. The letter has been published several times: see Angelo de Gubernatis, *Storia dei Viaggiatori Italiani nelle Indie Orientali* (Livorno, 1875); Fontoura da Costa, *Les déambulations du Rhinoceros de Modofar*, pp. 33–41; Ugo Serani, "La realtà virtuale nel Cinquecento: il rinoceronte di Dürer," in Maria José de Lancastre, Silvano Peloso, and Ugo Serani, eds., *E vós, Tágides minhas: Miscellanea in onore di Luciana Stegagno Picchio* (Viareggio-Lucca: Mauro Baroni, 1999), pp. 649–662, here pp. 652–655.

24. Thus Fontoura da Costa, *Les déambulations du Rhinoceros de Modofar*, p. 25, following José de Figueiredo. Artur Anselmo was also of this opinion; see Serani, "La realtà virtuale nel Cinquecento," p. 655.

25. The translation is taken from the British Museum's own description of the image (inv. no. SL, 5218.161).

26. Fontoura da Costa, *Les déambulations du Rhinoceros de Modofar*, p. 24.

27. For a thorough presentation of this argument, see Augustine Brannigan, *The Social Basis of Scientific Discoveries* (Cambridge: Cambridge University Press, 1981).

28. In the face of the vast bibliography on Dürer, we have based our account on two authoritative works. The first is Erwin Panofsky's *The Life and Art of Albrecht Dürer*, which he completed in 1943 after more than two decades of work on the artist, beginning with his book on the theoretical work of Dürer and continuing with various studies of the artist's attitude toward antiquity, the engraving *Melencolia*, and others. The other is by Joseph Koerner, one of today's leading experts on Dürer and on the Northern Renaissance: "Albrecht Dürer: A Sixteenth-Century Influenza," in Giulia Bartrum, ed., *Albrecht Dürer and His Legacy: The Graphic Work of a Renaissance Artist* (London: British Museum Press, 2002), pp. 18–39, which contains a full bibliography down to that year. His argument fits ours like a glove.

29. Concha Huidobro, *Durero y la edad de oro del grabado alemán* (Madrid: Electa, 1997); Fernando Checa, ed., *Durero y Cranach: Arte y humanismo en la Alemania del Renacimiento* (Madrid: Museo Thyssen-Bornemisza y Fundación Caja Madrid, 2007).

30. Raymond Klibansky, Erwin Panofsky, and Fritz Saxl, *Saturn and Melancholy: Studies in the History of Natural Philosophy, Religion, and Art* (London: Thomas Nelson & Sons, 1964).

31. See, for example, Dürer's *Helmet Visor (Design for a Silver Harness for Emperor Maximilian I)* from around 1517 in the Albertina in Vienna; *Three Studies of a Helmet* (ca. 1514) in the Louvre; or the armor in one of his masterpieces of engraving, *Knight, Death and Devil* (1513). The exhibition held in Madrid in 2007 included a good selection of pieces of metalwork, armor, drawings, and studies of metallic surfaces; for the excellent catalogue see Checa, *Durero y Cranach*.

32. The comparison between the arts par excellence—and not only in Leonardo—is between painting and poetry. On this inexhaustible theme see, for example, the pages entitled "Word and Image" in W. J. T. Mitchell, *Iconology: Image, Text, Ideology* (Chicago: University of Chicago Press, 1986), pp. 42–45.

33. Translation by J. M. Massing in J. A. Levenson, eds., *Circa 1492: Art in the Age of Exploration*, exhibit catalogue, National Gallery of Art, Washington, DC (New Haven: Yale University Press, 1991), p. 300.

34. On the performative character of language, the classic work—besides Aristotle, of course—is John L. Austin, *How to Do Things with Words* (Oxford: Clarendon Press, 1962).

35. Donald F. Lach, *Asia in the Making of Europe*, vol. 2, *A Century of Wonder, Book One: The Visual Arts* (Chicago: University of Chicago Press, 1970), p. 164; T. H. Clarke, *The Rhinoceros from Dürer to Stubbs, 1515–1799* (New York: Sotheby's, 1986), pp. 23–25. On Burgkmair, see Stephanie Leitch, "Burgkmair's *Peoples of Africa and India* and the Origins of Ethnography in Print," *Art Bulletin* 91, no. 2 (2009): 134–159; reissued in Stephanie Leitch, *Mapping Ethnography in Early Modern Germany: New Worlds in Print Culture* (New York: Palgrave Macmillian, 2010), pp. 63–99; Jean Michel Massing, "Hans Burgkmair's Depiction of Native Africans," *RES: Anthropology and Aesthetics* 27 (1995): 39–51; reissued in Jean Michel Massing, *Studies in Imagery*, vol. 2: *The World Discovered* (New York: Pindar Press, 2007), pp. 114–140.

36. I owe these philological remarks to Javier Igea, who was delving into a German etymological dictionary (F. Kluge, *Etymologisches Wörterbuch*) in an attempt to counter my argument with his profound knowledge of German language and culture. Claudia Swan, "Ad vivum, naer het leven," p. 356, records the Dutch formula "gheconterfeyt naer het leven." The classic on the subject is Peter Parshall, "Imago Contrafacta: Images and Facts in the Northern Renaissance," *Art History* 16, no. 4 (1993): 554–579. Much recent literature has been devoted to both terms, *contrafacta* and *ad vivum*. On the first one, see Kusukawa, *Picturing the Book of Nature*, pp. 8–19; Claudia Swan, "Counterfeit Chimeras: Early Modern Theories of the Imagination and the Work of Art," in Alina Payne, ed., *Vision and Its Instruments, c. 1350–1750: The Art of Seeing and Seeing as an Art* (State College: Pennsylvania State University Press, 2015), pp. 216–237. On the second one, see Noa Turel, "Living Pictures: Rereading 'au vif,' 1350–1550," *Gesta* 50, no. 2 (2011): 163–182; Boudewijn Bakker, "Au vif—naar 't leven—ad vivum: The Medieval Origin of a Humanist Concept," in Anton W. A. Boschloo et al., eds., *Aemulatio: Imitation, Emulation and Invention*

in Netherlandish Art from 1500 to 1800: Essays in Honor of Eric Jan Sluijter (Zwolle: Waanders, 2011), pp. 37–52.

37. By coincidence, while I was writing these pages and reading William J. T. Mitchell's book, my colleague Alicia Jiménez published her book *Imagines Hibridae: Una aproximación postcolonialista al estudio de las necrópolis de la Bética* (Madrid: CSIC, 2008), which opens with an excursus on the meaning of the word *imago* in the Roman world.

38. Walter Benjamin, "The Work of Art in the Age of Mechanical Reproduction," in *Illuminations: Essays and Reflections*, ed. and with an introduction by Hannah Arendt; trans. Harry Zohn (New York: Schocken Books, 1969), pp. 217–251. Benjamin mentions the woodcut at the start of this essay, although for the purpose of his argument it is, logically enough, the lithograph that interests him more.

39. Williams M. Ivins, *Prints and Visual Communication* (Cambridge: MIT Press, 1969). The first edition was published by Harvard in 1953. On the role of the print in modern science, see too the sumptuous catalogue edited by Susan Dackerman, *Prints and the Pursuit of Knowledge in Early Modern Europe* (New Haven: Yale University Press, 2011); the cover illustration is a detail from Dürer's woodcut of the rhinoceros.

40. Ivins discussed this basic problem of scientific representation. Others have joined in the debate as well; see Martin Kemp, "'The Mark of Truth': Looking and Learning in some Anatomical Illustrations from the Renaissance and the Eighteenth Century," in W. Bynum and R. Porter, eds., *Medicine and the Five Senses* (Cambridge: Cambridge University Press, 1993), pp. 85–121; Antonio de Pedro, *El diseño científico, siglos XV–XIX* (Madrid: Akal, 1999); Brian W. Ogilvie, "Image and Text in Natural History, 1500–1700," in W. Lefèvre, J. Renn, and U. Schoepflin, eds., *The Power of Images in Early Modern Science* (Basel: Birkhäuser Verlag, 2003), pp. 141–167; and Swan, *Ad vivum, naer het leven*.

41. Ivins, *Prints and Visual Communication*, p. 42.

42. Ernst Gombrich, *Art and Illusion* (Princeton: Princeton University Press, 1960), pp. 81–82. Pamela H. Smith and Paula Findlen, "Commerce and the Representation of Nature in Art and Science," in Smith and Findlen, *Merchants and Marvels*, pp. 1–25, start out from Dürer's engraving to introduce the theme of that collection of articles.

43. Once again, in the face of an extensive bibliography, we have selected two texts that are fundamental for tracing the representations of Dürer's rhinoceros in Europe: Clarke, *The Rhinoceros from Dürer to Stubbs*; and Lach, *Asia in the Making of Europe*. From the more recent literature, see Antonio Bernat and Tamás Sajo, "El poder de las imágenes: Notas para una rinocerontología," in Inocencio Galindo and José Vicente Martin, eds., *Imagen y conocimiento: Tradición artística e innovación tecnológica* (Valencia: Universidad Politécnica de Valencia, 2008), pp. 161–189.

44. On the duel see Clarke, *The Rhinoceros from Dürer to Stubbs*, pp. 155–163. For the traces left by Dürer's engraving see, in addition to the works listed above, Francis Joseph Cole, "The History of Albrecht Dürer's Rhinoceros in Zoological Literature," in E. Ashworth, ed., *Science, Medicine, and History: Essays on the Evolution of Scientific Thought and Medical Practice, Written in Honour of Charles Singer* (London: Oxford University Press, 1953) vol. I, pp. 337–356.

45. Lach, *Asia in the Making of Europe*, p. 165.

46. I am grateful to Sagrario López Poza for sending me a number of images of the rhinoceros among emblemata, in which our animal is usually associated with strength and vigilance. See too the excellent work that Antonio Bernat is posting on the web as a continuation of Bernat and Sajo, "El poder de las imágenes": http://www.emblematica .com/blog/studiolum.html.

47. The name Bada seems to be derived from a Portuguese adoption of a Malay word, *badoh*, from which the common Portuguese and Spanish form *abada* is derived (Lach, *Asia in the Making of Europe*, p. 168). Covarrubias records the form *bada*, "a very ferocious animal more commonly called rhinoceros," attributes its etymology to the allegiance of Indian to its Hebrew origin (all languages were derived from Hebrew), and comments at length on the *bada* that was "for a long time in Madrid." He also repeats some of the classical characteristics that we have already come across (its connection with the unicorn, its hostility toward the elephant, the medicinal properties of its horn) and one that we have passed over but which may have some interest in connection with Dürer: the rhinoceros was also associated with solitude and the retreat, in some form, into melancholy and solitary wandering. Sebastián de Covarrubias, *Tesoro de la lengua castellana o española* (Barcelona: Alta Fulla, 1993), pp. 181–182. In a similar

vein, Fray Juan de San Jerónimo describes it in the gardens of El Escorial in 1583: "It is an ugly, melancholy and sad animal, clad in a kind of armor, disagreeable and unknown," in *Memorias de Fray Juan de San Gerónimo*, in M. Salva and P. Sainz de Baranda, *Colección de documentos inéditos para la Historia de España* (Madrid: Imprenta de la Viuda de Calero, 1845), vol. 7, p. 369.

48. This was the villa of the royal scribe Juan de Vargas. I am grateful to Angel Tuninetti for giving me a copy of the interesting essay by Juan Mejía, *Rinocerontes colombianos: Mirada a unos animales en el arte* (Bogota: Alcaldía Mayor de Bogotá, 2004), pp. 59–62.

49. Bernat and Sajó, "El poder de las imágenes."

50. The image appears in *La gran enciclopedia china* of 1728. The iconographies created by Dürer and later by Philippe Galle and James Parson were well known in the Orient; see Clarke, *The Rhinoceros from Dürer to Stubbs*, pp. 163–166.

51. Ibid. For a review of the history and life of the species and its reflection in culture, see too Kelly Enright, *Rhinoceros* (London: Reaktion Books, 2008). L. C. Rookmaaker's *Bibliography of the Rhinoceros: An analysis of the Literature on the Recent Rhinoceroses in Culture, History and Biology* (Rotterdam: A. A. Balkema, 1983) is an indispensable aid. Rookmaaker is the editor of another useful resource, the Rhino Resource Center; see http://www.rhinoresource center.com/.

52. William Ashworth, "The Persistent Beast: Recurring Images in Early Zoology Illustration," in Ellenius, *The Natural Sciences*, pp. 46–66.

53. Koerner, "Albrecht Dürer," p. 21.

54. Ibid., p. 22, where Koerner comments on Dürer's aesthetic excursus.

4. Chimera

1. Manuel R. Trelles, "El padre fray Manuel de Torres," *Revista de la Biblioteca Pública de Buenos Aires*, vol. 4 (1882), pp. 441–448; Guillermo Furlong, *Naturalistas argentinos durante la dominación hispánica* (Buenos Aires: Huarpes, 1948), pp. 332–350; José María López Piñero and Thomas Glick, *El Megaterio de Bru y el presidente Jefferson: Una relación insospechada en los albores de la paleontología* (Valencia: Consejo Superior de Investigaciones Científicas [hereafter CSIC], 1993), pp. 39–43; Francisco Pelayo, *Del diluvio al megaterio. Los orígenes de la paleontología en España* (Madrid: CSIC, 1996),

pp. 294–302; Francisco de las Barras y de Aragón, "Un dibujo del Megaterio del Río Luxán," *Las ciencias* 11, no. 1 (1964): 77–85.

2. The correspondence is in the Biblioteca Nacional de la República Argentina (Ms. 5070) and has been reproduced several times, in, e.g., Trelles, "El padre fray Manuel de Torres," pp. 441–448; and Furlong, *Naturalistas argentinos durante la dominación hispánica*, pp. 332–350.

3. Loreto, April 30, 1787, quoted in Furlong, *Naturalistas argentinos durante la dominación hispánica*, p. 341.

4. José Custodio Sáa y Faria was an advisor to the viceroy on matters connected with the establishments on the Patagonian coast. He was also one of the main assistants to the Malaspina Expedition during its visit to Río de la Plata in 1789.

5. AGI (Archivo General de Indias, Seville), Mapas y planos de Buenos Aires, leg. 248 and 249. There is a copy in the Museo Nacional de Ciencias Naturales, Madrid (hereafter MNCN), *Expediente Megaterio*, caja no. 91. See María Ángeles Calatayud, *Catálogo crítico de los documentos del Real Gabinete de Historia Natural, 1787–1815* (Madrid: CSIC, 2000), ref. no. 64.

6. Loreto's comments are in the "Oficio del Marqués de Loreto, Virrey de Buenos Aires al excmo. Sr. D. Antonio Porlier . . . anunciando la remesa de la osamenta de un animal muy corpulento y desconocido," March 2, 1788, fol. 1 v°; AGI, Buenos Aires, leg. 76; copy in the MNCN, *Expediente Megaterio*, caja no. 91; Calatayud, *Catálogo crítico*.

7. The dimensions are those recorded by Sáa y Faria in the first of the drawings under discussion. The weight of the sacrum is taken from a third document accompanying the two drawings, an inventory giving the contents and weight of the seven crates in which the skeleton was transported: "Cuadro con número, peso, dimensiones y contenido en cada uno de los siete caxones que se remiten." AGI, Buenos Aires, leg. 249; copy in the MNCN, *Expediente Megaterio*, caja no. 91; Calatayud, *Catálogo crítico*.

8. José Custodio Sáa y Faria, *Copia del esqueleto de un Animal desconocido que se halló soterrado en la barranca del Río de Luxán*, AGI, Mapas y planos de Buenos Aires, leg. 248. Copy in the MNCN, *Expediente Megaterio*, caja no. 91; Calatayud, *Catálogo crítico*. A height of 18 feet is equivalent to 4.86 meters (1 foot = 27 cm). Unless otherwise indicated, translations into English from the original language are by Peter Mason.

9. Máximo Izzi, *Diccionario ilustrado de los monstruos. Ángeles, diablos, ogros, dragones, sirenas y otras criaturas del imaginario* (Barcelona: Olañeta, 2000), pp. 405–407.

10. On Torrubia, see Leandro Sequeiros and Francisco Pelayo, eds., *Aparato de Historia Natural Española de Joseph Torrubia* (Universidad de Granada, 2007). The introduction covers the theme of gigantology and the role of Torrubia's claims in the contexts of the debates of the time. See too López-Piñero and Glick, *El Megaterio de Bru y el presidente Jefferson*, pp. 27–33; Horacio Capel, *La física sagrada: Creencias religiosas y teorías científicas en los orígenes de la geomorfología española* (Barcelona: Ediciones del Serbal, 1985), pp. 51–59.

11. Irina Podgorny, "De ángeles, gigantes y megaterios: Saber, dinero y honor en el intercambio de fósiles de las provincias del Plata en la primera mitad del siglo XIX," in Ricardo D. Salvatore, ed., *Los lugares del saber: Contextos locales y redes transnacionales en la formación del conocimiento moderno* (Beatriz Buenos Aires: Viterbo, 2007), pp. 125–158, here p. 130.

12. Lorraine Daston and Katharine Park, *Wonders and the Order of Nature, 1150–1750* (New York: Zone Books, 1998); Peter G. Platt, ed., *Wonders, Marvels, and Monsters in Early Modern Culture* (Newark: University of Delaware Press, 1999); Antonio Lafuente and Javier Moscoso, eds., *Monstruos y seres imaginarios en la Biblioteca Nacional* (Madrid: Doce Calles, 2000); Peter Mason, *Before Disenchantment: Images of Exotic Animals and Plants in the Early Modern World* (London: Reaktion Books, 2009).

13. Daston and Park, *Wonders and the Order of Nature*, p. 350.

14. Ibid., pp. 109–133.

15. A good example is Juan Eusebio Nieremberg, *Historia Naturae. Maxime Peregrinae* (Antwerp: Plantijn, 1635).

16. Jorge Cañizares, *Católicos y puritanos en la colonización de América* (Madrid: Marcial Pons Historia, 2008), p. 187.

17. Michel Foucault, *The Order of Things: An Archaeology of the Human Sciences* (New York: Pantheon Books, 1971), p. 155.

18. "Oficio del Marqués de Loreto, Virrey de Buenos Aires al excmo. Sr. D. Antonio Porlier . . . anunciando la remesa de la osamenta de un animal muy corpulento y desconocido," March 2, 1788; AGI, Buenos Aires, leg. 76, f. 2; copy in the MNCN, *Expediente Megaterio*, caja no. 91; Calatayud, *Catálogo crítico*, ref. no. 64.

19. Antonio Porlier to the viceroy of Buenos Aires, September 2, 1788; Furlong, *Naturalistas argentinos durante la dominación hispánica*, p. 348.

20. *Carolus III Rex Naturam et Artem sub Uno Tecto in publicam utilitatem consociavit.* On the founding of the Royal Cabinet, see Agustín J. Barreiro, *El Museo de Ciencias Naturales (1771–1935)* (Madrid: Doce Calles, 1992 [1944]); Maria Ángeles Calatayud, *Pedro Franco Dávila y el Real Gabinete de Historia Natural* (Madrid: CSIC, 1988); Juan Pimentel, *Testigos del mundo: Ciencia, literatura y viajes en la Ilustración* (Madrid: Marcial Pons Historia, 2003), pp. 147–178; M. Villena, J. S. Almazán, J. Muñoz, and F. Yagüe, *El gabinete perdido: Pedro Franco Dávila y la Historia Natural del Siglo de las Luces*, 2 vols. (Madrid: CSIC, 2008).

21. Pedro Franco Dávila, Jean-Paul Gua de Malves, and Jean-Baptiste Romé de L'Isle, *Catalogue Systématique et Raisonné des curiosités de la Nature et de l'Art qui composent le cabinet de M. Davila, avec figures en taille-douce, de plusieurs morceaux qui n'avoient point encore éte gravés* . . . , 3 vols. (Paris: Briasson, Paris, 1767), vol 1, p. vii.

22. Cited in Barreiro, *El Museo de Ciencias Naturales*, p. 82.

23. Calatayud, *Catálogo crítico*, ref. no. 34.

24. Ibid., ref. no. 43.

25. López-Piñero and Glick, *El Megaterio de Bru y el presidente Jefferson*; José María López Piñero, "Juan Bautista Bru y la difusión por Cuvier de su obra paleontológica," *Arbor*, 527–528, 134 (1989): 79–99; López Piñero, *Juan Bautista Bru de Ramón: El atlas zoológico, el Megaterio y las técnicas de pesca valencianas, 1742–1799* (Valencia: Ayuntamiento de Valencia, 1996). In spite of the interest and importance of this scholar's work on Bru, on which I have drawn for these pages, I cannot share his opinion of the merits of the Valencian painter and taxidermist. López Piñero regrets that Cuvier in his day and Anglo-Saxon historians today (Alan Moorehead and Martin Rudwick) have not paid sufficient attention to Bru or have ignored him altogether, thereby reducing his role in the discovery of the Megatherium. This is to exaggerate the talent, training, and contribution of Bru and to turn it into one more episode in the polemic on Iberian science, or, what is worse, an example of an unjustly neglected tradition that he sets out to vindicate, namely, science in Valencia. Simply, my concerns have been different.

26. Joseph Garriga, *Descripción del esqueleto de un cuadrúpedo muy corpulento y raro, que se conserva en el Real gabinete de Historia Natural de Madrid* (Madrid, 1796). There are excellent reproductions of the plates in López Piñero, *El atlas zoológico*.

27. Libro de Cuentas del Real Gabinete de Historia Natural, caja no. 59, 1776–1809, pp. 115ff., MNCN.

28. Ibid., p. 120.

29. Mariano de la Paz Graells, "Estado de la conservación en que se halla el esqueleto del Megatherium Americanum, defectos que se notan en las piezas que le constituyen, corrección de estos y restauración o separación de las partes que en mi concepto faltan o sobran," one of the documents included in the *Memoria sobre el Megaterio del Museo de Madrid*, 1845, MNCN; Calatayud, *Catálogo crítico*, ref. no. 64: "Ever since the celebrated G. Cuvier published his report on the skeleton of the Megatherium preserved in our Museum of Natural History, this object has not ceased to attract the attention of naturalists abroad. . . . However, it is quite strange that regarding this curious creature that has been studied more from afar than at close quarters by eminent men of science, there are still various errors and differences of opinion, principally regarding the position and use of some of the parts of which it is composed. The origin of the former may lie in the mistakes committed by the taxidermist who assembled this skeleton and because an undeniable lack of the requisite anatomical knowledge marred the position of various bones. This is the origin not only of the striking deformities that have given this piece an extraordinary appearance but also of the mistaken ideas of those who, without a thorough examination of the work, have described the different parts as belonging to the place that they are already seen to occupy in the skeleton and in the plates that represent it. All the same, I must pay due respect to the great Cuvier who, with no other resources than the plates copied from the drawings made by Juan Bautista Bru and the scanty information that he was able to obtain, placed the Megatherium in the correct zoological niche and even emended or detected several of the errors that were made in assembling it." MNCN, *Expediente Megaterio*, caja no. 91. Equally critical words can be found in a report of August 23, 1833, by Juan Vilanova, another great naturalist and physician, and in his "Observaciones sobre el esqueleto del Megaterio

que se halla en el Gabinete de Historia Natural de Madrid," *Gazeta de Madrid* 35 no. 12 (1835): 1404.

30. Other museums of natural history and natural sciences that exhibit reconstructions of the same species (in London, Valencia, Amsterdam, and Argentina) have decided to represent it in an upright position, standing on its hind legs, with the front ones raised toward the trees.

31. Clavijo y Fajardo to Príncipe de la Paz, November 24, 1796, MNCN, *Expediente Megaterio*, caja no. 92. This correspondence concerns the plates and illustrations of the *Historia y Descripción de los Peces de nuestras costas* by Antonio Sáñez Reguart.

32. This and the following quotations are all taken from the *Descripción* published in 1796; see note 26.

33. First published in Latin as *Telluris Theoria Sacra* in 1681. See Paolo Rossi, *I segni del tempo. Storia Della Terra e storia delle nazioni da Hooke a Vico* (Milan: Feltrinelli, 1979), p. 58.

34. Ana V. Mazo, *Los cuatro elefantes del rey Carlos III* (Madrid: MNCN, 2008).

35. Ana V. Mazo, "El oso hormiguero de su majestad," *Asclepio* 58, no. 1 (2006): 281–294. There is a canvas in the MNCN representing this anteater, traditionally attributed to the workshop of Raphael Mengs and more recently to Goya.

36. See Marina Belozerskaya, *The Medici Giraffe and Other Tales of Exotic Animals and Power* (New York: Little, Brown, 2006), pp. 87–131. Carlos Gómez-Centurión, "Treasures Fit for a King: King Charles III of Spain's Indian Elephants," *Journal of the History of Collections* 22, no. 1 (2010): 29–44; Gómez-Centurión's *Alhajas para soberanos: Los animales reales en el siglo XVIII: De las leoneras a las mascotas de cámara* (Valladolid: Junta de Castilla y León, 2011) is an important monograph on the collecting of exotic animals in the Bourbon Spanish court.

37. Garriga, *Descripción*, pp. x–xi.

5. Bones

1. The former expression is attributed to Juan Bautista Bru in Joseph Garriga, *Descripción del esqueleto de un cuadrúpedo muy corpulento y raro, que se conserva en el Real gabinete de Historia Natural de Madrid* (Madrid, 1796), p. i (digital version available at http://aleph.csic.es/).

López Piñero attributes the latter to Carlos Gimbernat (1768–1834), geologist, mineralogist, and vice director of the Royal Cabinet, cited in José María López Piñero and Thomas Glick, *El Megaterio de Bru y el presidente Jefferson: Una relación insospechada en los albores de la paleontología* (Valencia: CSIC, 1993), p. 50.

2. Garriga, *Descripción*, p. ii.

3. Sebastian Serlio, *L'Architettura*, 7 vols. (1537–1575); Martin Kemp, "The Mark of Truth: Looking and Learning in Some Anatomical Illustrations from the Renaissance and the Eighteenth Century," in W. Bynum and R. Porter, eds., *Medicine and the Five Senses*, pp. 85–121 (Cambridge: Cambridge University Press, 1993).

4. Norton Wise, "Making Visible," *Isis* 97 (2006): 75–82.

5. For two good studies of Hooke with similar titles, see J. Bennett, M. Cooper, M. Hunter, and L. Jardine, *London's Leonardo: The Life and Work of Robert Hooke* (Oxford: Oxford University Press, 2003); Allan Chapman, *England's Leonardo: Robert Hooke (1635–1703) and the Art of Experiment in Restoration England* (Bristol: Institute of Physics, 2005).

6. Robert Hooke, *Micrographia: Some Physiological Descriptions of Minute Bodies Made by Magnifying Glasses with Observations and Inquiries Thereupon* (London, 1565), Preface.

7. Cited in Antonio de Pedro, *El diseño científico: Siglos XV–XIX* (Madrid: Akal, 1999), p. 22.

8. Cited in Kemp, "The Mark of Truth," p. 85.

9. Nuria Valverde, "Small Parts: Crisóstomo Martínez (1638–1694), Bone Histology, and the Visual Making of Body Wholeness," *Isis* 100, no. 3 (2009): 505–536.

10. Felipe Jerez, *Los artistas valencianos de la Ilustración y el grabado biológico y médico, 1759–1814* (Valencia: Ayuntamiento de Valencia, 1996), p. 418ff.

11. Ibid., p. 232ff.

12. J. Carrete Parrondo, F. Checa, and V. Bozal, *El grabado en España: Siglos XV–XVIII, Summa Artis*, vol. 31 (Madrid: Espasa, 2001); Juan Carrete Parrondo, *El grabado a buril en la España ilustrada: Manuel Salvador Carmona* (Madrid: Fábrica Nacional de Moneda y Timbre, 1989).

13. Antonio Gallego, *Catálogo de los dibujos de la Calcografía Nacional* (Madrid: Real Academia de Bellas Artes de San Fernando, 1978), p. 5.

14. For the biography of Bru and his other works see José María López Piñero, *Juan Bautista Bru de Ramón: El atlas zoológico, el Megaterio y las técnicas de pesca valencianas, 1742–1799* (Valencia: Ayuntamiento de Valencia, 1996); and Jerez, *Los artistas valencianos*.

15. López Piñero considered them possibly the best drawings by Bru. I cannot help thinking that they are too good to be his, although it is a complicated question of degree. An expert in vertebrate paleontology told me that she found the drawings full of errors.

16. The novel is *El dibujante de peces* (The fish draftsman) (Barcelona: Noray, 2007), by Juan Carlos Arbex, who attributed those drawings, among the best of their kind at the time, to a German soldier named Miguel Cros.

17. López Piñero, *Juan Bautista Bru de Ramón*, pp. 71–85 and 261–323; Jerez, *Los artistas valencianos*, pp. 278–289.

18. On Manuel Navarro, see Jerez, *Los artistas valencianos*, p. 323; and Elena Páez Ríos, *Repertorio de grabados españoles en la Biblioteca Nacional*, 4 vols. (Madrid: Ministerio de Cultura, 1985), vol. 2, pp. 280–282. On Salvador Carmona, see Carrete Parrondo, *El grabado a buril en la España ilustrada*.

19. Salvador Carmona's blindness was probably due to prolonged exposure to copper reflection. The blind conchologist is Georg Eberhard Rumphius (1627–1702). Problems of vision were frequent among the microscopists; the reason is easy to surmise: in the seventeenth and eighteenth centuries they carried out their observations practically in the dark or in very poorly illuminated conditions.

20. Carrete Parrondo, *El grabado a buril en la España ilustrada*, pp. 13–18, included his remarks on the manual on this technique by Manuel Rueda, *Instrucción para gravar en cobre y perfeccionarse en el gravado a buril, al agua fuerte, y al humo, con el nuevo método de grabar las planchas para estampar en colores, a imitación de la pintura; y un compendio histórico de los más célebres Gravadores, que se han conocido desde su invención hasta el presente* (Madrid: Imprenta de la viuda de Joaquín Ibarra, 1761).

21. The second plate contains the skull, two sections of a vertebra, and a clawed foot; the third the clavicle, sacrum, and a rib; and the fourth and fifth various bones belonging to the extremities and a tooth.

22. Dorinda Outram, *Georges Cuvier: Vocation, Science and Authority in Post-Revolutionary France* (London: Palgrave Macmillan, 1984).

23. On the Museum of Natural History in Paris, see Emma Spary, *Utopia's Garden: French Natural History from Old Regime to Revolution* (Chicago: University of Chicago Press, 2000).

24. "Notice sur le squelette d'une très-grande espèce de quadrupède inconnue . . ." *Magasin Encyclopédique* 2ᵉ année, 1 (1796): 303–310. It was the only work to be translated into English in the September 1796 issue of the *Monthly Magazine*, whose editors clearly had a flair for such items; it appeared there on pages 637–638. For fresh translations of this and other texts see Martin J. S. Rudwick, *George Cuvier, Fossil Bones, and Geological Catastrophes: New Translations & Interpretations of the Primary Texts* (Chicago: University of Chicago Press, 1997). There is a Spanish translation in Garriga, *Descripción*, pp. 17–20.

25. Cited in López Piñero and Glick, *El Megaterio de Bru y el presidente Jefferson*, p. 66.

26. Reproduction and translation in ibid., pp. 127–131. For the episode that led the Megatherium to brush against the doors of the White House, see ibid., pp. 72–89, and Julian P. Boyd, "The Megalonyx, the Megatherium and Thomas Jefferson's Lapse of Memory," *Proceedings of the American Philosophical Society* 102, no. 5 (October 1958): 420–435. When the Megalonyx appeared in 1796, Jefferson had forgotten what Carmichael had told him seven years earlier.

27. López Piñero and Glick assume that it must have been Bru, but we cannot exclude the possibility that it was Manuel Navarro or the surgeon from the Hospital del Buen Suceso.

28. Charles A. Miller, *Jefferson and Nature: An Interpretation* (Baltimore: Johns Hopkins University Press, 1988); I. Bernard Cohen, *Science and the Founding Fathers: Science in the Political Thought of Jefferson, Franklin, Adams, and Madison* (New York: Norton, 1995); Silvio Bedini, *Jefferson and Science* (Charlottesville: Thomas Jefferson Foundation, 2002).

29. Susan Stewart, "Death and Life, in That Order, in the Works of Charles Wilson Peale," in John Elsner and Roger Cardinal, eds., *The Cultures of Collecting* (London: Reaktion Books, 1994), pp. 204–224; Florike Egmond and Peter Mason, *The Mammoth and the Mouse: Microhistory and Morphology* (Baltimore: Johns Hopkins University Press, 1997), pp. 1–36; Lawrence Weschler, *Mr. Wilson's Cabinet of Wonder: Pronged Ants, Horned Humans, Mice on Toast, and Other Marvels of Jurassic Technology* (New York: Pantheon Books, 1995).

30. Bedini, *Jefferson and Science*, pp. 60–64.

31. Paul Semonin, *American Monster: How the Nation's First Creature Became a Symbol of National Identity* (New York: New York University Press, 2000).

32. Antonello Gerbi, *The Dispute of the New World: The History of a Polemic, 1750–1900*, rev. and enlarged ed., trans. Jeremy Moyle (Pittsburgh: University of Pittsburgh Press, 1973).

33. Bedini, *Jefferson and Science*, pp. 60–64; Semonin, *American Monster*, pp. 302–314.

34. Caspar Wistar, "A Description of the Bones Deposited, by the President, in the Museum of the Society, and Represented in the Annexed Plates," *Transactions of the American Philosophical Society* (1799), vol. 4, pp. 526–531; Thomas Jefferson, "A Memoir on the Discovery of Certain Bones of a Quadruped of the Clawed Kind in the Western Parts of Virginia," in ibid., pp. 246–260.

35. Comte de Buffon, *Histoire naturelle, générale et particulière* (1749–1788), vols. 5 and 13. See Gerbi, *The Dispute of the New World*; Fernando Ramírez and Irina Podgorny, "Las metamorfosis del Megaterio," *Ciencia hoy* 11, no. 61 (February/March 2001): 12–19.

36. Although it is customary to apply the term "debate on Iberian science" to the post-Restoration period, it is not difficult to find traces of it in the historiographical polemics of the eighteenth century, on which a good part of the late nineteenth-century debate was to draw for its attitude toward the singularity of Spain within the European context, its contribution to modern philosophy, or its emphasis on a Spanish national character.

6. Fossil

1. The scene described by Lyell is in "The Stinkstones of Oeningen," chapter 7 of Stephen Jay Gould, *Hen's Teeth and Horses' Toes: Further Reflections in Natural History* (New York: Norton, 1984), pp. 94–106.

2. For a political biography of Cuvier, see Dorinda Outram, *Georges Cuvier: Vocation, Science and Authority in Post-Revolutionary France* (Manchester: Manchester University Press, 1984).

3. His key works in both areas are: *Le règne animal distribué d'après son organisation pour servir de base à l'histoire naturelle des animaux et d'introduction à l'anatomie comparée*, 4 vols. (Paris, 1817); and *Recherches sur les ossemens fossiles de quadrupèdes où l'on rétablit les carac-*

tères de plusieurs espèces d'animaux que les révolutions du globe pa-roissent avoir détruites, 4 vols. (Paris, 1812). The latter includes his major contribution to geological history in the form of the "Prelimi-nary Discourse," which was subsequently published on its own on various occasions and in multiple languages *(Discours sur les révolu-tions de la surface du globe, et sur les changemens qu'elles ont produits dans le règne animal)*. We have consulted the 1836 edition of *Recher-ches* (Paris: Edmond d'Ocagne).

4. Martin J. S. Rudwick, *Bursting the Limits of Time: The Reconstruc-tion of Geohistory in the Age of Revolution* (Chicago: University of Chicago Press, 2005), a fundamental work for the writing of the present chapter. It is based on the Tarner Lectures delivered in Trinity College Cambridge in 1996, and was followed by a sequel, *Worlds before Adam: The Reconstruction of Geohistory in the Age of Reform* (Chicago: University of Chicago Press, 2008). These two vol-umes constitute a milestone in the history of paleontology, geology, and related sciences.

5. Cuvier, *Recherches* (1836 ed.), vol. 1, p. 80.

6. Michel Foucault, *The Order of Things: An Archaeology of the Human Sciences* (New York: Pantheon Books, 1971), p. 132.

7. Stephen Jay Gould, *Time's Arrow, Time's Cycle: Myth and Metaphor in the Discovery of Geological Time* (Cambridge: Harvard University Press, 1987).

8. Lorraine Daston and Katharine Park, *Wonders and the Order of Na-ture, 1150–1750* (New York: Zone Books, 1998).

9. Foucault, *The Order of Things*, p. 156.

10. Martin J. S. Rudwick, *The Meaning of Fossils: Episodes in the History of Palaeontology* (New York: Science History Publications, 1972), p. 24.

11. Ibid., p. 5ff; Brian W. Ogilvie, "Image and Text in Natural History, 1500–1700," in Wolfgang Lefévre, Jürgen Renn, and Urs Schoepflin, eds., *The Power of Images in Early Modern Science*, pp. 141–167 (Basel: Birkhäuser, 2003).

12. Giuseppe Olmi, *L'inventario del mondo: Catalogazione della natura e luoghi del sapere nella prima età moderna* (Bologna: Il Mulino, 1992), p. 165.

13. On Steno, see Alan Cutler, *The Seashell on the Mountaintop: A Story of Science, Sainthood, and the Humble Genius Who Discovered a New History of the Earth* (New York: Penguin, 2003).

14. Allan Chapman, *England's Leonardo: Robert Hooke (1635–1703) and the Art of Experiment in Restoration England* (Bristol: Institute of Physics, 2005). Leonardo da Vinci, by the way, also took an interest in fossils.

15. *The Posthumous Works of Robert Hooke, M.D. S.R.S. . . . Containing his Cutlerian Lectures, and other Discourses* (London: Smith and Walford, 1705), p. 335; Rudwick, *Bursting the Limits of Time*, p. 194.

16. There is a vast literature on the subject. Besides Rudwick, *The Meaning of Fossils*, we have drawn mainly on Paolo Rossi, *The Dark Abyss of Time: The History of the Earth and the History of Nations from Hooke to Vico* (Chicago: University of Chicago Press, 1984), originally published as *I segni del tempo: Storia della Terra e delle Nazioni da Hooke a Vico* (Milan: Giangiacomo Feltrinelli Editore, 1979); Horacio Capel, *La física sagrada: Creencias religiosas y teorías científicas en los orígenes de la geomorfología española* (Barcelona: El Serbal, 1985); and Francisco Pelayo, *Del Diluvio al Megaterio: Los orígenes de la paleontología en España* (Madrid: CSIC, 1996). In spite of the titles, the two books in Spanish are not confined to theories of the Flood and of paleontology on the Iberian Peninsula alone in the early modern period.

17. This is the title of Capel's book, *La física sagrada*. It is taken from Johann Jakob Scheuchzer, who used it for his *Jobi Physica Sacra* (1721) and later for his best-known work, *Physica Sacra* (1731–1735).

18. Rudwick, *The Meaning of Fossils*, p. 71ff.

19. On Mosaic chronology and the history of the earth, see Rossi, *The Dark Abyss of Time*, and Capel, *La física sagrada*, p. 28ff.

20. Capel, *La física sagrada*, pp. 42–69.

21. Foucault, *The Order of Things*, p. 132.

22. Genesis 6:4: "There were giants in the earth in those days."

23. José Luis Sanz, *Cazadores de dragones: Historia del descubrimiento e investigación de los dinosaurios* (Barcelona: Ariel, 2007), p. 32.

24. *Mundus subterraneus, quo universae denique naturae divitiae* (Rome, 1664). See Paula Findlen, ed., *Athanasius Kircher: The Last Man Who Knew Everything* (New York: Routledge, 2004).

25. Rudwick, *The Meaning of Fossils*, p. 78ff.; Rossi, *The Dark Abyss of Time*, pp. 33–41; Capel, *La física sagrada*, pp. 106–111; Gould, *Time's Arrow*, chap. 2.

26. Christopher Hill, *The World Turned Upside Down: Radical Ideas during the English Revolution* (London: Maurice Temple Smith, 1972), pp. 19–38.

27. See above, note 17.

28. For theories about the Flood and the intermingling of sacred and profane history, see Rossi, *The Dark Abyss of Time*, p. 217ff.

29. Gottfried W. Leibniz, *Protogaea*, trans. Claudine Cohen and Andre Wakefield (Chicago: University of Chicago Press, 2008).

30. Arthur J. Lovejoy, *The Great Chain of Being* (Cambridge: Harvard University Press, 1936).

31. On his geological ideas, see Georges-Louis Leclerc, Comte de Buffon, *Las épocas de la naturaleza*, ed. Antonio Beltrán (Madrid: Alianza, 1997). For the dialectic between reason and sentiment, science and literature, see the introductory study in *Georges-Louis de Leclerc, Conde de Buffon 1707–1788*, ed. Antonio Lafuente and Javier Moscoso (Madrid: CSIC, 1999), pp. ix–lxxx. See also Jacques Roger, *Buffon: Un philosophe au Jardin du Roi* (Paris: Fayard, 1989); Thierry Hoquet, *Buffon: Histoire naturelle et philosophie* (Paris: Honoré Champion, 2005).

32. Gould, *Time's Arrow*, chap. 1.

33. Ibid., chap. 3; Rudwick, *Bursting the Limits of Time*, pp. 158–172.

34. This work on the external characteristics of fossils was translated twenty years later by Andrés Manuel del Río for the Mining Seminar of Mexico using an old term for mineralogy that has now fallen into disuse, oryctology: *Elementos de Orictognosia o del conocimiento de los fósiles, dispuestos según los principios de A. G. Werner, para el uso del Real Seminario de Minería de México* (1795). This was the first work of its kind to be published in New Spain and was acclaimed by Alexander von Humboldt.

35. Rudwick, *Bursting the Limits of Time*, p. 197ff.

36. Ibid., p. 234.

37. It was published in Gotha in *Magazin für den neuste aus der Physic und Naturgeschichte* 6, no. 4: 117ff. See Rudwick, *Bursting the Limits of Time*, p. 295ff.

38. Ibid., pp. 181–194; Rossi, *The Dark Abyss of Time*, p. 36ff.; Gould, *Time's Arrow*, pp. 3–8.

39. On archaeology in Enlightenment Spain, see Gloria Mora, *Historias de mármol: La arqueología clásica española en el siglo XVIII* (Madrid: CSIC, 1998).

40. There is a vast literature on the discovery of Pompeii and Hercula-neum. Besides ibid., pp. 108–112, the most detailed monograph in Spanish is Félix Fernández Muga, *Carlos III y el descubrimiento de Herculano, Pompeya y Estabia* (Salamanca: Ediciones Universidad de Salamanca, 1989). More wide-ranging still is Alain Schnapp, *The Discovery of the Past: The Origins of Archaeology* (London: British Museum Press, 1996). Rudwick, *Bursting the Limits of Time*, p. 187ff., uses Schnapp with great discretion to construct his argument, although with a certain disregard for (and lack of knowledge about) the policy of King Charles VII. On this point Rudwick allows himself to be carried away by the Anglo-Saxon cliché in claiming that Charles tried to keep the discoveries secret. Of course, for Rudwick it was Hamilton who was the hero who publicized the discoveries; he naturally fails to mention Alcubierre. This classic interpretation has provoked the inevitable violent rejoinders; it is hard to know which is worse—the stereotypes of the black legend or some overly patriotic responses.

41. Gould, *Time's Arrow*, pp. 10–16.

42. Bettina Dietz and Thomas Nutz, "Collections Curieuses: The Aesthetics of Curiosity and Elite Lifestyle in Eighteenth-Century Paris," *Eighteenth-Century Life* 29, no. 3 (2005): 44–74.

43. Rudwick, *Bursting the Limits of Time*, p. 239ff.

44. Peter Dance, *A History of Shell Collecting* (Leiden: Brill, 1986).

45. Joseph Garriga, *Descripción del esqueleto de un cuadrúpedo muy corpulento y raro, que se conserva en el Real gabinete de Historia Natural de Madrid* (Madrid, 1796), p. xiv.

46. The former claim is made, and the latter suggested, in José María López Piñero and Thomas Glick, *El Megaterio de Bru y el presidente Jefferson: Una relación insospechada en los albores de la paleontología* (Valencia: CSIC, 1993).

47. Robert P. W. Visser, *The Zoological Work of Petrus Camper (1722–1789)* (Amsterdam: Rodop, 1985).

48. Rudwick, *Bursting the Limits of Time*, p. 263ff.; Peter Simon Pallas, *Reisen durch verschiedene Provinzen des Russischen Reichs*, 4 vols. (St. Petersburg, 1771–1776).

49. Claudine Cohen, *Le destin du mammouth* (Paris: Seuil, 1994).

50. Rudwick, *Bursting the Limits of Time*, p. 274.

51. William Hunter, "Observations on the bones, commonly supposed to be elephants bones which have been found near the river Ohio in America," *Philosophical Transactions* 58 (1768): 34–45.

52. Paul Semonin, *American Monster: How the Nation's First Creature Became a Symbol of National Identity* (New York: New York University Press, 2000).

53. See, for example, Carolyn Gilman, *Lewis and Clark: Across the Divide* (Washington, DC: Smithsonian Books and Missouri Historical Society Press, 2003).

54. Emma Spary, *Utopia's Garden: French Natural History from Old Regime to Revolution* (Chicago: University of Chicago Press, 2000).

55. William B. Ashworth, "The Persistent Beast: Recurring Images in Early Zoological Illustration," in Allan Ellenius, ed., *The Natural Sciences and the Arts* (Uppsala: Almqvist & Wikell, 1985), pp. 46–66, here p. 57ff.; Peter Mason, *Before Disenchanment: Images of Exotic Animals and Plants in the Early Modern World* (London: Reaktion Books, 2009), pp. 15–19 and 200–206.

56. Georges-Louis Leclerc, Comte de Buffon, *Histoire naturelle, générale et particulière* (Paris, 1749–1788), vols. 5 and 13.

57. Georges Cuvier, "Notice sur le squelette d'une très-grande espèce de quadrupède inconnue jusqu'à present, trouvé au Paraguay, et déposé au Cabinet d'Histoire Naturelle de Madrid," *Magasin Encyclopédique*, 2ᵉ année, 1 (1796): 303–310.

58. Georges Cuvier, "Mémoire sur les espèces d'éléphans tant vivantes que fossiles," *Magasin Encyclopédique*, 2ᵉ année, 3 (1796): 440–445. English translation: Martin J. S. Rudwick, *George Cuvier, Fossil Bones, and Geological Catastrophes: New Translations & Interpretations of the Primary Texts* (Chicago: University of Chicago Press, 1997), pp. 18–24. Cuvier published an extended version three years later in *Mémoires de l'Institut National des Sciences et des Arts, sciences mathématiques et physiques* 2 (1799), pp. 1–22, pls. 2–6.

59. The 1993 film *Jurassic Park*, based on the novel *The Lost World* by Michael Crichton.

60. Stéphane Schmitt, *Aux origines de la biologie moderne: L'anatomie comparée d'Aristote à la théorie de l'évolution* (Paris: Éditions Belin, 2006).

61. Rudwick, *Bursting the Limits of Time*, pp. 89–97.

62. Ibid.

63. Outram, *Georges Cuvier*; Dorinda Outram, "New Spaces in Natural History," in N. Jardine, J. A. Secord, and E. C. Spary, eds., *Cultures of Natural History* (Cambridge: Cambridge University Press, 1996), pp. 249–266.

64. Rudwick, *Bursting the Limits of Time*, p. 363.

65. Georges Cuvier, "Antiquaire d'une espèce nouvelle, il me fallut apprendre à la fois à restaurer ces monumens des révolutions passés et à en déchiffrer le sens," *Discours sur les révolutions de la surface du globe, et sur les changemens qu'elles ont produits dans le règne animal,* in *Recherches sur les ossemens fossiles* (1836 ed.), vol. 1, pp. 93–414, here p. 93; English translation of the 1812 edition with commentary is by Rudwick, *George Cuvier*, pp. 173–252.

66. Rudwick, *Bursting the Limits of Time*, p. 369.

67. Cuvier, *Recherches* (1836 ed.), vol. 1, pp. 96–97; Rudwick, *George Cuvier*, p. 185.

68. On the representation of the skull in the Iberian world, see Fernando Rodríguez de la Flor, *La península metafísica* (Madrid: Biblioteca Nueva, 1999), p. 216ff.

69. Georges Cuvier, "Extrait d'un Mémoire sur un animal dont on trouve les ossements dans la pierre à plâtre des environs de Paris, & qui paraît ne plus exister vivant aujourd'hui," delivered in the Museum of Natural History on October 6, 1798. English translation: Rudwick, *George Cuvier*, pp. 35–41; original French text: pp. 285–290.

70. Martin J. S. Rudwick, *Scenes from Deep Time: Early Pictorial Representations of the Prehistoric World* (Chicago: University of Chicago Press, 1992).

71. Semonin, *American Monster*, pp. 341–348; Florike Egmond and Peter Mason, *The Mammoth and the Mouse: Microhistory and Morphology* (Baltimore: Johns Hopkins University Press, 1997), pp. 1–18.

72. See Margaret Meredith, "Friendship and Knowledge: Correspondence and Communication in Nothern-Transatlantic Natural History, 1780–1815," in Simon Schaffer, Lissa Roberts, Kapil Raj, and James Delbourgo, eds., *The Brokered World: Go-Betweens and Global Intelligence, 1770–1820* (Uppsala: Studies in History of Science, 2009), pp. 151–191.

73. Pietro Corsi, *The Age of Lamarck: Evolutionary Theories in France, 1790–1830* (Berkeley: University of California Press, 1988). This is an

expanded and revised edition of his *Oltre il mito: Lamarck e le scienze naturali del suo tempo* (Bologna: Il Mulino, 1983).

74. Rudwick, *Bursting the Limits of Time*, pp. 394–396.

75. On the scientific aspects of the Napoleonic campaign in Egypt, see Charles C. Gillispie, *Science and Polity in France: The Revolutionary and Napoleonic Years* (Princeton: Princeton University Press, 2004); Fernand Beaucour, Yves Laissus, and Chantal Orgogozo, *La découverte de l'Egypte* (Paris: Flammarion, 1990); Marie-Noëlle Bourguet and Yves Laissus, *Il y a 200 ans les savants en Égypte* (Paris: Muséum d'histoire naturelle/Nathan, 1998); Maria Luisa Ortega, "Ciencia y civilización: La expedición de Bonaparte y el Egipto moderno," Ph.D. thesis (Universidad Autónoma de Madrid, 1997).

76. Georges Cuvier, "Sur le Megatherium, autre animal de la famille des paresseux, mais de la taille du rhinocéros, dont un squelette fossile presque complet est conservé au Cabinet Royal d'Histoire Naturelle à Madrid," *Annales du Muséum d'Histoire Naturelle*, 5 (1804), pp. 376–400; reprinted in his *Recherches sur les ossements fossiles* of 1812; see *Recherches* (1836 ed.), vol. 8, pp. 331–370.

77. Fernando Ramírez and Irina Podgorny, "Las metamorfosis del Megaterio," *Ciencia hoy* 11, no. 61 (February/March 2001): 12–19.

78. Juan Pimentel, "Across Nations and Ages: The Creole Collector and the Many Lives of the Megatherium," in Schaffer, Roberts, Raj, and Delbourgo, *The Brokered World*, pp. 321–353.

79. Cuvier, "Sur le Megatherium," p. 29; *Recherches* (1836 ed.), vol. 8, p. 363.

80. Cited in Irina Podgorny, "De ángeles, gigantes y megaterios: Saber, dinero y honor en el intercambio de fósiles en las provincias del Plata en la primera mitad del Siglo XIX," in Ricardo Salvatore, ed., *Los lugares del saber: Contextos locales y redes transnacionales en la formación del conocimiento moderno*, pp. 125–158, here p. 128 (Rosario: Beatriz Viterbo, 2007).

81. H. C. Pander and J. d'Alton, *Die vergleichende Osteologie: 1. Das Riesenfaulthier, Bradypus giganteus, beschrieben, und mit den verwandten Geschlechtern verglichen* (Bonn: Weber, 1821). D'Alton had also illustrated Pander's work on embryology in *Beiträge zur Entwicklungsgeschichte des Hühnchens im Eye* (Würzburg: Brönner, 1817).

82. H. B. Nisbet, *Goethe and the Scientific Tradition* (London: Institute of Germanic Studies, 1972); George A. Wells, *Goethe and the Development of Science, 1750–1900* (Alphen aan den Rijn: Sijthoff and

Noordhoff, 1978); Frederick Amrine, Francis J. Zucker, and Harvey Wheeler, eds., *Goethe and the Sciences: A Reappraisal*, Boston Studies in the Philosophy of Science (Dordrecht: Springer, 1987); R. H. Stephenson, *Goethe's Conception of Knowledge and Science* (Edinburgh: University of Edinburgh Press, 1995); Robert J. Richards, *The Romantic Conception of Life: Science and Philosophy in the Age of Goethe* (Chicago: University of Chicago Press, 2002); Roger H. Stephenson, "Binary Synthesis: Goethe's Aesthetic Intuition in Literature and Science," *Science in Context* 18, no. 4 (2005): 553–581.

83. Johann Wolfgang von Goethe, *The Metamorphosis of Plants*, trans. Douglas Miller (Cambridge: MIT Press, 2009). See too Juan Pimentel, "Entre el imperio vegetal y el sagrado enigma: Linneo, Goethe y el lenguaje de las plantas," in R. Olmos, P. Cabrera, and S. Montero, eds., *Paraíso cerrado, jardín abierto: El reino vegetal en el imaginario religioso del Mediterráneo* (Madrid: Polifemo, 2005), pp. 297–320.

84. Although he arrived at different conclusions, in the sixth edition of *The Origin of Species* Darwin recognized the priority of Goethe in certain ideas on the proximity and evolution of animals and living forms (he claimed that Goethe had played a role in Germany similar to the one played by his grandfather Erasmus Darwin in England and Geoffroy Saint-Hilaire in France before the century drew to a close and the work of Lamarck arrived on the scene). Besides his theses on the primordial plant *(Urpflanze)*, Goethe had discovered the presence of the intermaxillary bone in man, which had been supposed to be nonexistent in the human species.

85. Richard Owen, "On the Megatherium (Megatherium Americanum, Cuvier and Blumenbach)," "Part 2, Vertebrae of the Trunk," *Philosophical Transactions of the Royal Society of London* 145 (1855): 359–388, here p. 360, where Owen relates the history of research on the Megatherium and the deductions and contributions of his predecessors. For the versions of the Megatherium of Goethe, Pander, and d'Alton, see too Wells, *Goethe and the Development of Science*, pp. 36–46; Richards, *The Romantic Conception of Life*, pp. 478–484.

86. Podgorny, "De ángeles, gigantes y megaterios," p. 133. See too Stéphane Schmitt, "From Eggs to Fossils: Epigenesis and Transformation of Species in Pander's Biology," *International Journal of Developmental Biology* 49, no. 1 (2005): 1–8.

87. Joseph Edward d'Alton, *Naturgeschichte des Pferdes* (Bonn, 1810–1816); later Rudolph Kuntz and Joseph Edward d'Alton, *Abbildungen saemmtlicher Pferde-Raçen* (Bonn, 1827).

88. Rudwick, *Scenes from Deep Time*; Jane P. Davidson, *A History of Paleontology Illustration* (Bloomington: Indiana University Press, 2008). Specifically on vertebrates, see Mauricio Antón, "La reconstrucción de vertebrados fósiles," in Inocencio Galindo and José Vicente Martín, eds., *Imagen y conocimiento. Tradición artística e innovación tecnológica* (Valencia: Universidad Politécnica de Valencia, 2008), pp. 65–77.

89. Ramírez and Podgorny, "Las metamorfosis del Megaterio"; Schmitt, *Aux origines de la biologie moderne.*

90. It was sent by Dámaso Antonio Larrañaga, a presbyter from Montevideo, a figure somewhere between Manuel de Torres and Jefferson (as he played a role during the struggle for independence), and launched a debate that was to go on for two decades. See Podgorny, "De ángeles, gigantes y megaterios," p. 134.

91. Nicolas A. Rupke, *The Great Chain of History: William Buckland and the English School of Geology, 1814–1849* (Oxford: Oxford University Press, 1983); Rudwick, *Bursting the Limits of Time,* p. 600ff.

92. *The Bridgewater Treatises: On the Power Wisdom and Goodness of God as Manifested in the Creation,* 8 vols. (1833–1840). See J. Topham, "Beyond the 'Common Context': The Production and Reading of the Bridgewater Treatises," *Isis* 89 (1998): 233–262.

93. William Buckland, *Geology and Mineralogy Considered with Reference to Natural Theology,* 2nd ed., 2 vols. (London: William Pickering, 1837), vol. 1, p. 144.

94. Podgorny, "De ángeles, gigantes y megaterios."

95. William Clift, "Notice on the Megatherium Brought from Buenos Aires by Woodbine Parish, Esq. FRS," *Transactions of the Geological Society,* 2nd ser., 3 (1835): 435–460.

96. Richard Owen, "On the Megatherium (Megatherium Americanum, Cuvier and Blumenbach)," "Part 2, Vertebrae of the Trunk," *Philosophical Transactions of the Royal Society of London* 145 (1855): 359–388, here p. 362.

97. Henri Marie Ducrotay de Blainville, *Ostéographie ou description iconographique comparée du squelette et du système dentaire des mammifères récents et fossiles* (1839–64).

98. Richard Owen, "Description of a tooth and part of the skeleton of the Glyptodon, a large quadruped of the edentate order, to which belongs the tessellated bony armor figured by Mr. Clift in his memoir on the remains of the Megatherium, brought to England by Sir Woodbine Parish, F.G.S.," *Proceedings of the Geological Society of London* (1839) 3, pp. 108–113; Richard Owen, *Description of the Skeleton of an Extinct Gigantic Sloth, Mylodon robustus Owen, with Observations on the Osteology, Natural Affinities, and Probable Habits of the Megatherioid Quadrupeds in General* (London: R. & J. Taylor, 1842).

99. Richard Owen, "On the Megatherium (Megatherium americanum, Cuvier and Blumenbach)," "Part 1, Preliminary Observations on the Exogenus Processes of Vertebrae," *Philosophical Transactions of the Royal Society of London (PTRSL)* 141 (1851): 719–764; "Part II, Vertebrae of the Trunk," *PTRSL* 145 (1855): 359–388; "Part III, The Skull," *PTRSL* 146 (1856): 571–589; "Part IV, Bones of the Anterior Extremities," *PTRSL* 148 (1858): 261–278; "Part V, Bones of the Posterior Extremities," *PTRSL* 149 (1859), pp. 809–829. See too Richard Owen, *Zoology of the Voyage of the Beagle* (1840), "Part 1, Fossil Mammalia," pp. 57–106.

100. R. D. Keynes, ed., *Charles Darwin's Beagle Diary* (Cambridge: Cambridge University Press, 1998), p. 107.

101. Charles Darwin, *The Voyage of the Beagle* (Ware: Wordsworth, 1997), pp. 79–81.

102. There is an immense literature on dinosaurs. See, for example, Deborah Cadbury, *The Dinosaur Hunters* (London: Fourth Estate, 2000); and Sanz, *Cazadores de dragones*.

103. To follow the role of the Megatherium in the early theories of speciation and in the Victorian parables, there is an unpublished thesis available at academia.edu: Anna Miriam Toledano, *The Posthumous Lives of the Giant Sloth: The Megatherium's Path from Artifact to Idea* (B.A. thesis, History Department, Princeton University, 2011).

104. Darwin, *The Voyage of the Beagle*, p. 81.

Epilogue

1. Peter Sloterdijk, *Globes: Macrospherology II: Spheres* (Los Angeles: Semiotexte/Smart Art, 2014).

2. On the opposition between experimental facts and writing—that is, between the books of the world and the world of books—see Hans Blumenberg, *Die Lesbarkeit der Welt* (Frankfurt: Suhrkamp, 1979.

3. Bruno Latour, *Science in Action* (Cambridge: Harvard University Press, 1987); Bruno Latour, *Pandora's Hope* (Cambridge: Harvard University Press, 1999).

4. Walter Benjamin, "The Work of Art in the Age of Mechanical Reproduction," in *Illuminations: Essays and Reflections*, ed. and with an introduction by Hannah Arendt; trans. Harry Zohn (New York: Schocken Books, 1969), pp. 217–251.

5. John M. Ulrich, "Thomas Carlyle, Richard Owen, and the Paleontological Articulation of the Past," *Journal of Victorian Culture* 11, no. 1 (2006): 30–58.

ACKNOWLEDGMENTS

In some way or other, what began as an address to a seminar ended up as an obsession that swallowed up other projects and held me in its thrall for several years. When my friends asked me what I was working on (the equivalent to "Where do you live?") and I told them about my extravagant analogy between the rhinoceros and the Megatherium, some of them smiled and, a few months later, asked me: "How are your camels?"—a normal reaction. Others displayed more empathy in giving me postcards or magnets with the rhinoceros. There were also those who shared in my concerns, helped me to think about the problem, and gave me information.

Fernando Bouza recommended that I read a work on another strange couple, *The Mammoth and the Mouse*. Although the reflections of Florike Egmond and Peter Mason on microhistory and morphology exceeded my essay, and their filiation with Carlo Ginzburg and structuralism lay outside my immediate concerns, at a conference organized by another friend and historian of science, José Pardo Tomás, I had the opportunity to meet Peter Mason, with whom I have collaborated on various projects and who has become a friend. He was the ideal translator for this book.

Various aspects of this work have been presented at seminars and conferences: Susana Gómez invited me to the Universidad Complutense de Madrid, Tiago Saraiva to the Instituto de Ciencias Sociais de Lisboa; Bartolomé Yun asked me to speak at the European University Institute in Florence, and Stéphane van Damme to do the same at SciencesPo in Paris. I am grateful to Stéphane for the encouragement and insistence on having the book translated into English. It was he who put me in contact with Ian Malcolm, an editor at Harvard University Press, the kind of professional any writer is fortunate to meet. It was a pleasure working with him. Helen Cowie, a bilingual colleague, helped us with the final corrections.

The flexibility of the Megatherium enabled me to adapt it to a project on go-betweens and the circulation of knowledge, in which I benefited from the wisdom of Simon Schaffer, my friend and mentor since my years in Cambridge. Since then, Lissa Roberts has been another admired colleague and frequent discussion partner. James Delbourgo and Kapil Raj accompanied me on that adventure, as did Neil Safier on that one and on many others too. He is another of those who urged that this book should be translated. Isabel Soler and Julio Pardos were its first readers. The list of colleagues from whom I have benefited is enormous. It includes José Ramón Marcaida, Sandra Sáenz López, Fernando Rodríguez de la Flor, María Tausiet, Felipe Pereda, José Pardo Tomás, Antonio Lafuente, Leoncio López-Ocón, Alberto Corsín, Nuria Valverde, Paco Pelayo, Vanessa de Cruz, Irina Podgorny, Daniela Bleichmar, Gloria Mora, Henique Leitão, and Antonella Romano. I was assisted in the Museo de Ciencias Naturales in Madrid, where the Megatherium resides, by Carmen Velasco, Isabel Morón, and Begoña Sánchez Chillón. Although my intellectual debts are acknowledged in the notes, some deserve to be highlighted here, such as T. H. Clarke's splendid book on the rhinoceros, the classic lucid work of William Ivins on the print and visual communication, the research of José María López Piñero on the Megatherium, and above all the monumental *Bursting the Limits of Time* by Martin S. J. Rudwick. Years ago I was able to attend some of his Tarner Lectures in Trinity College, Cambridge, the very lectures that triggered the genesis of his book.

The Fundación Jorge Juan in Madrid, headed by José de la Sota, funded the translation into English. Fernando Guerrero, editor of Abada Editores, which published the original edition in Spanish, has been more than an editor for this book and for its author. I must also mention the two research projects of the Spanish Ministry of Science under whose aegis the English version was forged, "Naturalezas figuradas" (HAR2010-15099) and "Imágenes y fantasmas de la ciencia ibérica" (HAR2014-52157-P).

My mother managed to read the book before she died, and Silvia accompanied and supported me as ever. But I would like to dedicate this book to two other pairs of fantastic creatures: my children, Carmen and Javier, and my brother and sister, Paco and Aurora.

CREDITS

Frontispiece. See Figure 54.

Figure 1. Belem Tower and its gargoyle. Pen and ink drawings by Jorrín Montañés. Courtesy of the artist.

Figure 2. Detail of *Rhinocerus 1515*. Woodcut by Albrecht Dürer.

Figure 3. Hanno, Pope Leo X's elephant, attributed to Raphael or Giulio Romano (after a lost drawing by Raphael). (Photo: Wikimedia Commons, CC-PD-Mark)

Figure 4. Hanno epitaph, by Francisco de Holanda, sketchbook, 1538. El Escorial Library, Madrid. (Photo: Wikimedia Commons, CC-PD-Mark)

Figure 5. Terreiro do Paço, detail of Lisbon view, in Georg Braun and Franz Hogenberg, *Civitates Orbis Terrarum*, 1582. (Photo: Biblioteca Nacional de España, Madrid, GMG/433)

Figure 6. Detail of the tapestry cycle, *The Lady and the Unicorn* (*La Dame à la Licorne*, Cluny Abbey, Paris).

Figure 7. The fight between the rhinoceros and the elephant, in Ambroise Paré, *Discourse de la mumie, de la licorne, des venins et de la peste* (1582). (Photo: Biblioteca Pública de Lyon)

Figure 8. Albrecht Dürer, *The Large Piece of Turf*, 1503 (Albertina, Vienna). (Photo: Wikimedia Commons, CC-PD-Mark)

Figure 9. Dürer, *Rhinoceron 1515*. Pen and sepia ink drawing. (Bequeathed by Sir Hans Sloane. SL,5218.161. Photo: © The Trustees of the British Museum)

Figure 10. Dürer, *Rhinocerus 1515*, woodcut. (Photo: National Gallery of Art, Washington, DC, Rosenwald Collection, 1964.8.697)

Figure 11. Hans Burgkmair, *Rhinoceros 1515*, woodcut. (Graphische Sammlung, Albertina, Vienna). (Photo: Wikimedia Commons, CC-PD-Mark)

Figure 12. David Kandel, *Rhinoceros*, engraving in Sebastian Münster, *Cosmographiae*, 1544.

Figure 13. Conrad Gessner, *Historia animalium*, 1551–1558.

Figure 14. Zoomorphic columns, in Joseph Boillot, *Nouveaux portraits et figures des termes pour user en* Architecture, 1592.

Figure 15. Emblem of Alessandro de Medici, from Paolo Giovio's *Dialogo dell'impresse militari et amorose*, 1555. (Photo: Wikimedia Commons, CC-PD-Mark)

Figure 16. José Custodio Sáa y Faria, Partes del esqueleto con sus dimensiones. (Photo: Courtesy of Ministerio de Educación, Cultura y Deporte. Archivo General de Indias, Seville, AGI, MP-Buenos Aires,249)

Figure 17. José Custodio Sáa y Faria, *Copia del esqueleto de un Animal desconocido que se halló soterrado en la barranca del Río de Luxán.* (Photo: Courtesy of Ministerio de Educación, Cultura y Deporte. Archivo General de Indias, Seville, AGI, MP-Buenos Aires, 248)

Figure 18. Chimera of Arezzo, Etruscan bronze *(above)* (Museum of Archeology, Florence), and a reproduction of it in a public fountain in Arezzo *(below)* (unidentified photographer)

Figure 19. Juan Bautista Bru and Manuel Navarro, plate number 1, published in José Garriga, *Descripción del esqueleto de un cuadrúpedo muy corpulento y raro, que se conserva en el Real gabinete de Historia Natural de Madrid* (1796). (Photo: Courtesy Museo Nacional de Ciencias Naturales, Madrid, ACN0091/064)

Figure 20. Andrea Vesalius, *De Humani Corporis Fabrica* (1543). (Photo: U.S. National Library of Medicine, Bethesda, MD, 2295005R)

Figure 21. Giulio Casserio, *Theatrum anatomicum, in Anatomische Tafeln* (1656), Frontispiece. (Photo: U.S. National Library of Medicine, Bethesda, MD. Dream Anatomy exhibition, History of Medicine Division)

Figure 22. William Cheselden, *Osteographia or the Anatomy of Bones* (1733). (Photo: U.S. National Library of Medicine, Bethesda, MD. WZ 260 C4990 1733 OV2)

Figure 23. Bernard Siegfried Albinus, *Tabulae sceleti et musculorum corporis humani* (1747). (Photo: U.S. National Library of Medicine, Bethesda, MD. WZ 260 A337ts 1749 OV2. Reproduced from the 1749 edition)

Figure 24. *Encyclopédie's* frontispiece, drawing by Charles-Nicolas Cochin in 1764, and engraving by Benoît Louis Prévost in 1772.

Figure 25. Juan Bautista Bru, preparatory drawing for plate number 1 in José Garriga, *Descripción del esqueleto de un cuadrúpedo muy corpulento y raro, que se conserva en el Real gabinete de Historia Natural de Madrid* (1796) (see Figure 19). (Photo: Courtesy Museo Nacional de Ciencias Naturales, Madrid, ACN110B/005/0064)

Figure 26. Juan Bautista Bru and Manuel Navarro, detail of plate number 2, published in José Garriga, *Descripción del esqueleto de un cuadrúpedo . . .* (1796). (Photo: Courtesy Museo Nacional de Ciencias Naturales, Madrid, ACN11OB/005/0064)

Figure 27. Juan Bautista Bru and Manuel Navarro, plate number 2, published in José Garriga, *Descripción del esqueleto de un cuadrúpedo . . .* (1796). (Photo: Courtesy Museo Nacional de Ciencias Naturales, Madrid, ACN11 OB/005/0064)

Figure 28. Georges Cuvier, *Megatherium americanum*, plate published in "Notice sur le squelette . . . ," *Magasin Encyclopédique*, 2e année, 1, 1796.

Figure 29. Georges Cuvier, Skulls of a two-toed sloth, a three-toed sloth, and Megatherium, illustration published in "Notice sur le squelette . . . ," *Magasin Encyclopédique*, 2e année, 1, 1796.

Figure 30. Caspar Wistar, fossils of claws of Megalonyx, "A Description of the Bones Deposited, by the President, in the Museum of the Society, and Represented in the Annexed Plates," *Transactions of the American Philosophical Society*, vol. 4, 1799.

Figure 31. Detail of one of the earliest images of a cabinet of curiosities, engraving in Ferrante Imperato, *Dell'Historia Naturale* (1599).

Figure 32. Ammonites drawn by Robert Hooke and described in his *Discourses of Earthquakes and Subterraneous Eruptions*, published after his death in 1705.

Figure 33. Thomas Burnet, *Telluris Theoria Sacra* (1681). Frontispiece reproduced from second edition (1689).

Figure 34. Coins and fossils. The usual analogy between these two objects considered both to be traces or remains of (natural or social) past forms. (*Bottom*): Georg Wolfgang Knorr, Johann Ernst Immanuel Walch, ed. *Die Naturgeschichte der Versteinerungen* (1768–1773).

Figure 35. "Veduta interior o atrio d'un Tempio nella parte occidentale di Pozzuolo," in Paolo Antonio Paoli, *Antichità di Pozzuoli* (Naples, 1768).

Figure 36. A Lamarckian plate, in W. I. May, *An Illustrated Index of Tasmanian Shells* (1923). (Photo: University of Tasmania Special and Rare Materials Collections)

Figure 37. Maastricht Animal jaws, in Faujas de Saint Fond, *Histoire Naturelle de la Montagne de Saint-Pierre de Maestricht* (1799).

Figure 38. The beast named, successively, Ohio Animal, American *incognitum*, American Mammoth, and Mastodon. Drawing sent by

Everard Home to Georges Cuvier, 1804. (Photo: © Muséum National d'Histoire Naturelle, Bibliothèque Centrale, La Direction Nationale des Bibliothèques et de la Documentation, Paris. Ms 630)

Figure 39. Georges- Louis Leclerc, Comte de Buffon, *Histoire Naturelle, générale et particulière* (1749–1788).

Figure 40. Skeleton of a terrestrial salamander. Its reconstruction and that of the Montmartre marsupial caused similar problems. Georges Cuvier, *Recherches sur les Ossemens Fossiles, Fourth Edition,* Atlas (1836), Vol. 2, Plate 254.

Figure 41. *Vanitas.* Human skull *(above),* in Govert Bidloo, *Anatomia Humani Corporis* (1685) (Photo: U.S. National Library of Medicine, Bethesda, MD. 2312021R. Reproduced from the 1690 Dutch edition), and skull of the Megatherium *(below),* by Bru and Navarro, in José Garriga, *Descripción del esqueleto de un cuadrúpedo . . .* (1796). (Photo: Courtesy Museo Nacional de Ciencias Naturales, Madrid, ACN11OB/005/0064)

Figure 42. Georges Cuvier, *Megatherium fossile,* in *Recherches sur les Ossemens Fossiles . . .* (1812 ed.), Vol. 4.

Figure 43. Two sketches by Cuvier, in which he shows his gift for the animated restoration of the big extinct vertebrates. Anoplotherium Commune *(above)* (Photo: © Muséum National d'Histoire Naturelle, Bibliothèque Centrale, Paris. Ms 635) and Palaeotherium Minus *(below), Recherches sur les Ossemens Fossiles, Fourth Edition,* Atlas (1836), Vol. 1, Plate 146. The former in the Museum of Natural History (Paris) and the latter from a copy in the University Library (Cambridge).

Figure 44. The studio of Benjamin Water house Hawkins in Sydenham, where he made the Crystal Palace dinosaurs. (Photo: Wikimedia Commons, CC-PD-Mark)

Figure 45. H. C. Pander and J. d'Alton, *Das Riesen-Faulthier, Bradypus giganteus* (1821).

Figure 46. The Megatherium, in H. C. Pander and J. d'Alton, *Das Riesen-Faulthier* (1821) *(above)*, and Georges Cuvier, *Recherches sur les Ossemens Fossiles* . . . (1836 ed.) *(below)*

Figure 47. Georges Cuvier, *Recherches sur les Ossemens Fossiles* . . . Atlas (1836 ed.), Vol. 2, Plate 217.

Figure 48. The Megatherium according to J. d'Alton, in William Buckland, *Geology and Mineralogy Considered with Reference to Natural Theology*, Vol. 2 (1836). (Photo: Courtesy Museo Nacional de Ciencias Naturales, Madrid)

Figure 49. William Buckland, *Geology and Mineralogy Considered with Reference to Natural Theology*, Vol. 2 (1836). (Photo: Courtesy Museo Nacional de Ciencias Naturales, Madrid)

Figure 50. *Above:* Glyptodon, Heinrich Harder (1858–1935) (Photo: Wikimedia Commons, CC-PD-Mark). *Below:* Glyptodon, *Museo civico di Storia Naturale*, Milan. (Photo: © Giovanni Dall'Orto. Wikimedia Commons, CC-BY-3.0)

Figure 51. The Iguanodon dinner party, New Year's Eve, 1853. Notice the plates with names of Buckland, Cuvier, Owen, and Mantell. *The Illustrated Encyclopaedia of Dinosaurs* (1854). (Photo: Wikimedia Commons, CC-PD-Mark)

Figure 52. Henry A. Ward, *Notice of the Megatherium Cuvieri* (1864). Frontispiece.

Figure 53. Megatherium at the Natural History Museum, London. (Photo: © Ballista. Wikimedia Commons, CC-BY-3.0)

Figure 54. The two primary images of this book: Dürer's woodcut (1515, see Figure 10) and Bru and Navarro's engraving (1796, see Figure 19).

Figure 55. The "negative" or the reverse of the previous images: the skeleton of the rhinoceros drawn by Cuvier, in *Recherches* (Vol. 2, 1812), and the

Megatherium with its fur, drawn by Mauricio Antón (courtesy of the artist).

Figure 56. The Megatherium from the Río Luján. (Photo: Courtesy Museo Nacional de Ciencias Naturales, Madrid)

INDEX

Note: Page numbers in *italics* indicate illustrations.